Petrochemicals
for the
Nontechnical Person

Petrochemicals
for the
Nontechnical Person

Donald L. Burdick

William L. Leffler

PennWell Books
PennWell Publishing Company
Tulsa, Oklahoma

Copyright © 1983 by
PennWell Publishing Company
1421 South Sheridan Road/P.O. Box 1260
Tulsa, Oklahoma 74101

Library of Congress cataloging in publication data

Burdick, Donald L.
    Petrochemicals for the nontechnical person.

    Includes index.
    1. Petroleum chemicals. I. Leffler, William L.
II. Title.
TP692.3.B87   1983   661'.804   82-16665
ISBN 0-87814-207-X

Printed in the United States of America

1   2   3   4   5   87   86   85   84   83

CONTENTS

PREFACE

Into the valley of death rode the three hundred.

"The Charge of the Light Brigade,"
Rudyard Kipling (1809–1892)

It is our observation that if you're not a chemist or a chemical engineer, you've probably had a hard time finding out what petrochemicals really are and how they're made. If you are a salesman, you've probably been to a couple of training courses because you must know something about the products you sell. But even there, the reading materials were probably too technical, dry, deadly.

This book is the second in a series designed to remedy the situation. The first is *Petroleum Refining for the Nontechnical Person*. You'd probably profit (even more than us) if you read that book before this one. The first two chapters are particularly important—on the nature of petroleum and on distillation. We're taking for granted in this book that you have a pretty good understanding of those two concepts.

You may be happy to know that this book is not quantitatively oriented, despite the technical nature of the subject. We concentrated on concepts and jargon. So this book should be particularly suited to the curious, aspiring novice as well as to the frustrated, seasoned veteran. It will be helpful for persons in finance, distribution, sales, public relations, advertising, planning, or purchasing areas of petrochemical, petroleum-based, or service companies.

As in *Petroleum Refining*, we've tried to outline the reading matter to let you use the book in several ways. We included a good index, so you will have a quick reference. We also made each chapter self-contained. For example, you can read about cumene and phenol without reading about propylene and benzene first. Finally, we designed the chapters with a commercial approach. If we plotted volume produced versus chapter number, the line would be steeply down to the right, except for the chapters on polymers at the end.

We kept this book small, so it can still be considered a primer and not an encyclopedia. Consequently, we limited the number of petrochemicals treated in detail. Probably everyone's favorite chemical has gotten the short shrift or has not been covered at all. We're sorry. If we were alchemists we might have avoided that problem. But we're a chemist and a businessman. That combination is our only apology.

We did all of the work ourselves, so we have no one to thank or blame except our secretaries, Beverly Mancuso and Debbie Huval. To encourage you to go on, we pass on the best review we could have gotten. It came from Debbie after she typed several of our toughest chapters. She said, "I probably wouldn't have dropped out of high school chemistry if you had written the book." We hope you'll stay for the full course.

I.

THE COMPLETE COURSE IN ORGANIC CHEMISTRY

"The time has come," the Walrus said, "to talk of many things: Of shoes—and ships—and sealing wax—of cabbages—and kings."

Through the Looking Glass,
Lewis Carroll (1832–1898)

What is *organic* chemistry? It's the study of compounds containing carbon, and it is fundamental to understanding petrochemicals. Why the word organic? Originally (and that means before 1800) organic applied only to compounds whose formation was supposedly due to some living force, such as plants or animals. Then early in the nineteenth century, a chemist named Wöhler synthesized urea, the main ingredient in urine. (History didn't record why he was trying to do that.) Until that time, it was believed urea could only be produced organically, by animal life. Thereafter, and until today, the term organic chemistry was stretched beyond its original meaning to include all carbon compounds, and the difference between organic and inorganic chemistry is more definitional than natural.

Surprisingly, organic compounds comprise more than 95% of all compounds known to exist, and that's over a million. Three characteristics of carbon and carbon compounds help explain the proliferation of organic chemicals. The first is the electronic configuration of the carbon atom. Don't leave now. You're about to get the 200-word summary of the *Periodic Table of Elements*, atoms, electrons, protons, valences, bonds, and compounds.

There are about 100 different atoms that make up all kinds of matter, and they can be classified in a table (the *Periodic Table of Elements*), according to how they are formed. In the center of any atom is a nucleus containing protons and neutrons. The protons have a positive electrical charge; the neutrons are neutral. Therefore, the nucleus is positively charged. Electrons (equal in number and charge to the protons) move around the nucleus in orbits. You might think of an atom like the solar system. The nucleus acts like the sun; the electrons orbit the nucleus like the planets circle the sun.

However, there is an important difference. The inner-most orbit can contain either one or two electrons. The next orbit can have up to eight electrons. The succeeding orbits become a more complex story, but luckily almost all petrochemicals have atoms with no more than two orbits.

The rules of electrons and orbits are important because the number of electrons in the outermost orbit largely determines the chemical properties of the element. Atoms strive for maximum stability by filling up their outer-most orbit. They can gain, shed, or share electrons with another atom in order to make a complete orbit of two or eight electrons. For example, consider the carbon atom. It has six neutrons and six protons in the nucleus and six electrons in orbit. The first orbit must have two; only four are left for the outer-most orbit. These four are called the *valence* electrons. Carbon has a valence of four because it needs four more electrons to fill the outer orbit. It desperately wants to find some other atoms with which it can share electrons.

Hydrogen also can be used as another example to complete the story. Hydrogen has only one proton in the nucleus and one electron in the first orbit, the valence electron. It needs another electron in that orbit to stabilize itself. Fig. 1–1 shows how carbon and hydrogen can achieve peace with each other. Each of four hydrogen atoms share their one electron with the carbon to satisfy carbon's need for eight electrons in the outer ring. The carbon shares an electron with each hydrogen to satisfy the need for two electrons in hydrogen's outer ring. The result is methane, CH_4, a compound that is chemically stable with all of the proton and electron charges balanced.

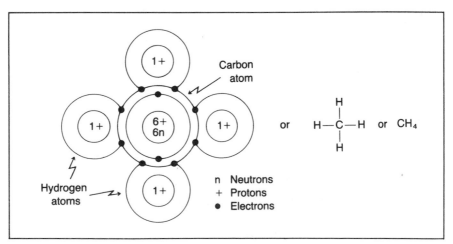

Fig. 1–1 Different representations of methane

Carbon and hydrogen can even link up with other carbons and hydrogens. When hydrogen hooks up with another hydrogen, it forms H_2 and the electron needs of each of the atoms are satisfied. But when carbon hooks up with another carbon, each carbon atom still has three unsatisfied electrons. Filling them out with hydrogens is one possibility. When that happens, the compound ethane forms, as shown in Fig. 1–2.

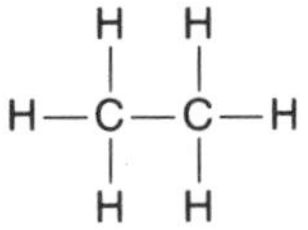

$$\begin{array}{cc}
\text{H} & \text{H} \\
| & | \\
\text{H}-\text{C}-\text{C}-\text{H} \\
| & | \\
\text{H} & \text{H}
\end{array} \qquad C_2H_6$$

Fig. 1–2 Ethane

The propensity of carbon to connect with four other atoms partially explains why so many carbon compounds exist. There are lots of other ways atoms can hook up with carbon, which we will discuss later.

The second characteristic unique to carbon compounds is called *isomerism* (from the Greek, *iso*, meaning "the same," and *meros*, meaning "parts"). Compounds with the same number and kind of atoms can have very different properties. For example, glucose has the formula $C_6H_{12}O_6$. Yet sixteen other compounds have that same number of carbon, hydrogen, and oxygen atoms. It's not likely, though, that you would want your nurse to hook up galactose or fructose to your intravenal instead of glucose, even though they both have the same formula. The difference is that the atoms are linked together in such a way as to have different spatial configurations. Such compounds are called isomers.

So if you put the phenomenon of isomerism together with the propensity of carbon to react (the valence of four) and add to that Mother Nature's blessing of an abundance of carbon in this universe, you can understand the preponderance of carbon compounds and the importance of organic chemistry.

One further characteristic unique to carbon is important and should be covered before leaving the subject of valences: bonds. As we discussed, carbon can link up to itself via a single bond with three other atoms. In fact carbon also can link to itself with double or triple bonds to satisfy its valence requirements of four. For example in Fig. 1–3, two carbon atoms are linked together with single, double, or triple bonds and are filled out with hydrogens, forming three different compounds.

This multiple-bond configuration is ironic. You will find that in petrochemical processes the greater the number of multiple bonds, the more reactive the compound generally is. Acetylene is more likely to react with other chemicals (explosively, in fact) than ethylene, which is far more reactive than ethane. You can think of multiple bonds as

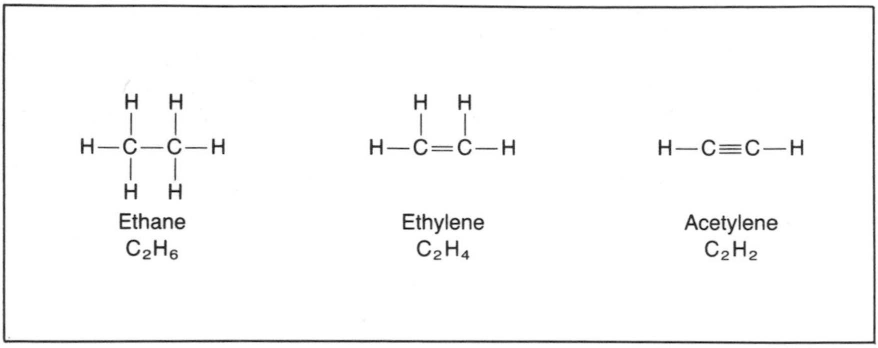

H   H
| |
H—C—C—H
| |
H   H
Ethane
C_2H_6

H   H
| |
H—C=C—H

Ethylene
C_2H_4

H—C≡C—H

Acetylene
C_2H_2

Fig. 1–3

squeezing into a place suitable for only one bond, hence increasing the reactiveness.

As a matter of common nomenclature in the petrochemical world, carbon compounds with single bonds are called *saturates*—the carbon atoms are saturated with other atoms. Those with multiple bonds are called *unsaturates*. Double-bond reactivity is characteristic of the basic building blocks in the petrochemical business. Ethylene, for example, is the chemical compound used to make vinyl chloride, ethylene oxide, ethyl alcohol, polyethylene, and styrene.

Going on with this study can be treacherous without a road map, so you need to look at one of the generally accepted breakdowns of organic chemicals shown in Fig. 1–4. The aliphatic hydrocarbons have already been introduced. They contain only hydrogen and carbon atoms, and they can have single or multiple bonds. Therefore, they are saturated or unsaturated hydrocarbons. The simplest member of the group is methane, and more complicated molecules in this group can be formed

$$\text{by adding} \quad \begin{matrix} \text{H} \\ | \\ -\text{C}- \\ | \\ \text{H} \end{matrix} \quad \text{between any carbon and hydrogen atoms, as shown}$$

in Fig. 1–5.

The family that results by continuing to add —CH_2 is called the *paraffin* series. This name comes from the wax used in the "old days" to seal jelly jars. That particular paraffin consists of a mixture of $C_{30}H_{62}$ on up to $C_{50}H_{102}$.

Unsaturated hydrocarbons are typified by ethylene. Propylene, butylene, and larger molecules are formed in the same manner as the saturates. However, one of the single bonds between the carbons is

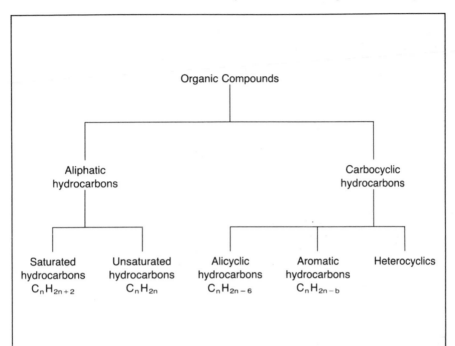

Fig. 1–4   Classification of organic chemicals

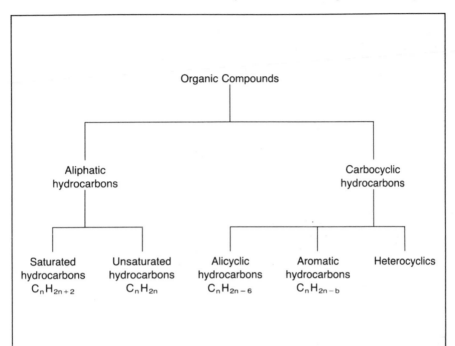

Fig. 1–5   Paraffins

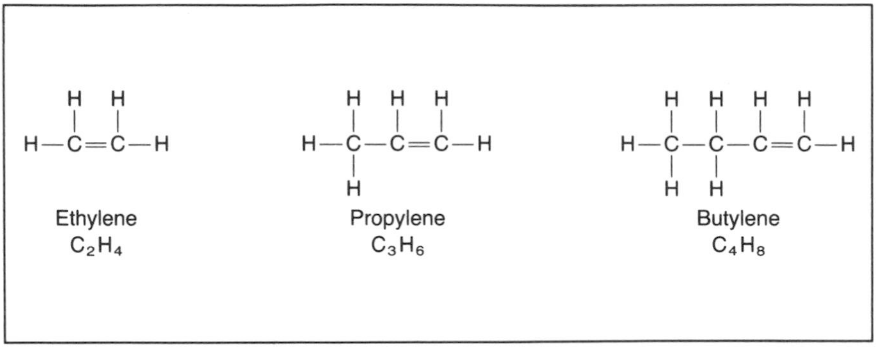

Ethylene C_2H_4 — Propylene C_3H_6 — Butylene C_4H_8

Fig. 1-6 Olefins

replaced with a double bond, as shown in Fig. 1–6. Another more popular name for these compounds is *olefins.*

The double-bond difference between olefins and paraffins is the essence of the difference between petrochemicals and petroleum: the former depend much more on the chemical reactivity of the double-bonded molecules. While paraffins can be manipulated in refineries by separation or reshaping, olefins in a petrochemical plant are usually reacted with another kind of atom or compound such as chlorine, oxygen, water, ammonia, or even with more of themselves. The result is more complicated compounds that are useful in an increasing number of chemical applications.

Two auxiliary, but important, concepts should be discussed. The first is isomers; the second is *organic groups.*

In order to understand isomers better, consider butane and its isomer, isobutane, and the three butylene isomers. The difference between the two C_4H_{10} molecules in Fig. 1–7 is how the carbons are connected to each other. In isobutane, one atom has three carbons attached to it, not two as in normal butane. However, the two butane molecules are drawn differently, and they behave differently. They boil at separate temperatures, give off individual amounts of heat when burned, have different chemical reactivity, and so on.

Butylene isomers are similar to butane isomers. However, they offer more options because the double bond destroys the symmetry. It's an easy mistake to go overboard drawing isomers that appear to have the same formula but look different. What looks different on paper may be identical when rolled over in space. Therefore, the isobutylene can be drawn at an angle. As long as one of the carbons has three other carbons attached to it, they remain in the same configuration no matter how you crank them around.

Normal butane
(C_4H_{10}, but sometimes
nC_4 for short)

Isobutane
(also C_4H_{10}, but
sometimes iC_4
for short)

Fig. 1–7 Butanes

Normal butylene
or
Butene-1
(C_4H_8 or $nC_4^{=}$)

Normal butylene
or
Butene-2
(C_4H_8 or $nC_4^{=}$)

Isobutylene
or
Isobutene
(C_4H_8 or $iC_4^{=}$)

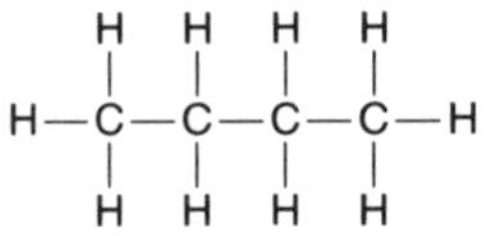

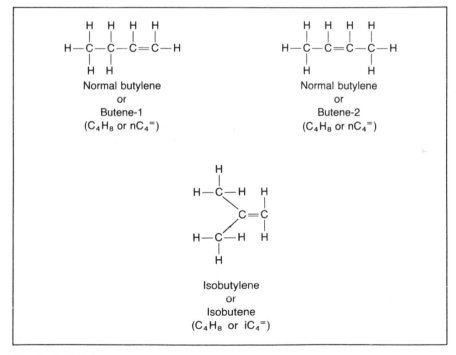

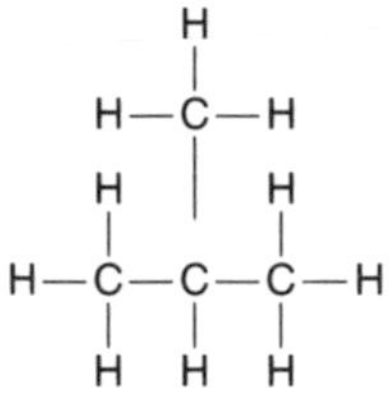

Fig. 1–8 Butylenes

Like the butane isomers, the butylenes each have their own properties that make them unique. Therefore, each has an individual appeal in the petrochemical industry.

Organic groups refer to groups of atoms that are a part of a molecule but look almost like the stand-alone molecule for which they are named. For example, take the methyl group shown in Fig. 1–9. It is a methane molecule missing one hydrogen, but it is attached to some other atoms to make a larger molecule, methyl alcohol. Organic groups are not stand-alone molecules themselves. They are always part of a molecule.*

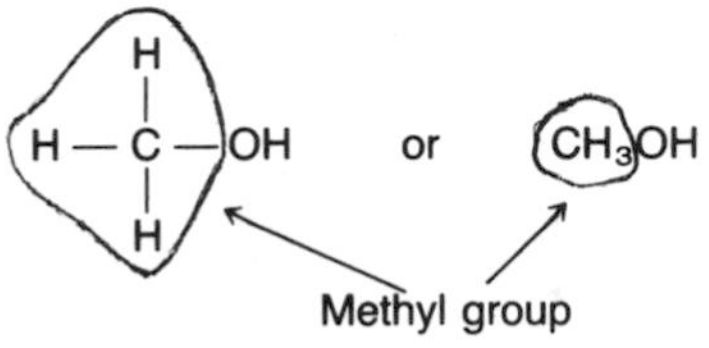

Methyl group

Fig. 1–9

The chemical shorthand for organic groups is R—. This notation is used when the writer wants to indicate that any of a number of organic groups could be attached to the molecule.

Another example of organic groups is shown in Fig. 1–10, a diagram of tetraethyl lead. This is the additive put in gasoline to improve the octane rating. Tetraethyl lead has four ethyl groups attached to lead (Pb).

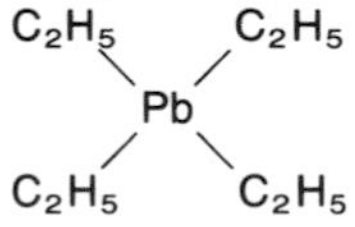

Fig. 1–10   Tetraethyl lead

---

*Often the terms *organic group* and *radical* are incorrectly used interchangeably. A radical looks like an organic group, except it can stand alone unattached to a molecule. As a result it has an unpaired, odd electron, and it is extremely reactive. The methyl radical,

$$H-\overset{\overset{\displaystyle H}{|}}{\underset{\underset{\displaystyle H}{|}}{C}}\cdot$$

, can be produced from methane by the loss of one hydrogen atom (H •). The use of the notation R, which is the first letter of the word radical, for organic group doesn't help dispel the confusion between the two terms.

Carbocyclic Compounds

Break down the word *carbocyclic* and you have the fundamental difference between these compounds and the others already discussed: carbon in a cyclic arrangement. These compounds have a closed chain of carbon atoms. Cyclopropane, shown in Fig. 1–11, is the simplest carbocyclic in the *alicyclic* group. (They are called alicyclic because for the most part compounds making up this group have properties like those of the aliphatic hydrocarbons already described.) You may have been administered cyclopropane in a dentist's chair—it's one of several anesthetics used. Others include laughing gas (nitrous oxide) and ether.

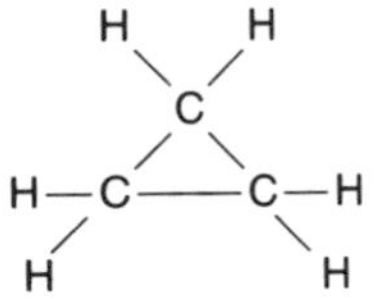

Fig. 1–11   Cyclopropane

Lower members of the alicyclic series have chemical properties similar to double-bonded olefins; they are quick to react. The simple explanation for this reactivity is that the bonds attaching the carbons to each other are really strained by the angles they must take. They are bent out of their natural shape. In any chemical reaction, the rings readily open up to alleviate this strain.

You might surmise, and correctly so, that if there are more carbon atoms in the ring, the compound is more stable. In fact, cyclopentane and cyclohexane (Fig. 1–12) are much more stable and, like the paraffins, are slow to react. They burn, of course, but not explosively. These two alicyclics among others are commonly found in petroleum products like gasoline and are generically called *naphthenes* in the refining business.

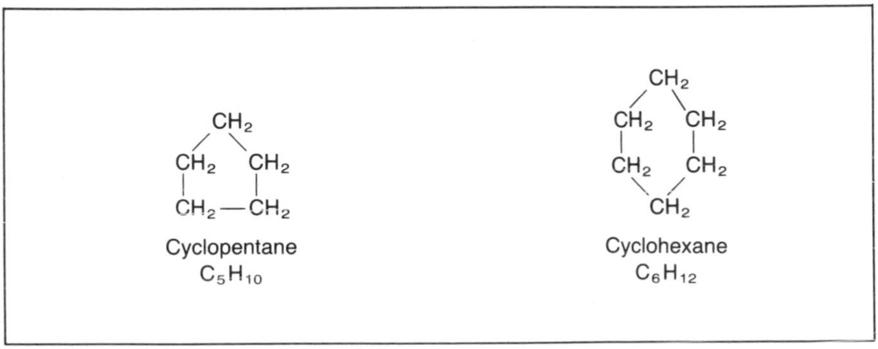

Cyclopentane
C_5H_{10}

Cyclohexane
C_6H_{12}

Fig. 1–12   Alicyclics

Aromatic Compounds

The most commercially important branch of compounds on the right side of the organic-compound family tree in Fig. 1–4 are the *aromatic* compounds. *Benzene* is the patriarch. Like much of the nomenclature in organic chemistry, the term "aromatic" is a misnomer. It is a legacy from the early nineteenth century when a group of unsaturated compounds differing in olefinic reactivity and with a sickly sweet, hydrocarbonish smell were isolated under that name. Unlike the term organic (which became broader in meaning), the name aromatics got narrower. Today it is limited to benzene and benzene derivatives. The latter term, benzene derivatives, is certainly more apropos than aromatics.

The benzene molecule is a remarkable structure. Six carbons are in a hexagonal ring, and every other carbon-to-carbon link is a double bond. To satisfy the valence rules, each carbon has only one hydrogen attached.

Some subtle but important characteristics are unique to the benzene ring. One is symmetry. Every carbon atom in the ring looks like every other carbon, and every hydrogen looks like every other hydrogen. Every benzene molecule also looks alike, but there are no benzene isomers. Furthermore, when one of the hydrogens is replaced in a chemical reaction (resulting in something called a *monosubstituted* benzene), that compound has no isomers.

For example take the compound toluene. In Fig. 1–14, all three molecules are benzene with a methyl group replacing a hydrogen. While they may appear to be oriented differently, each needs to be rotated just a little to look like the others. Therefore, only one kind of toluene actually exists, and it is a monosubstituted benzene.

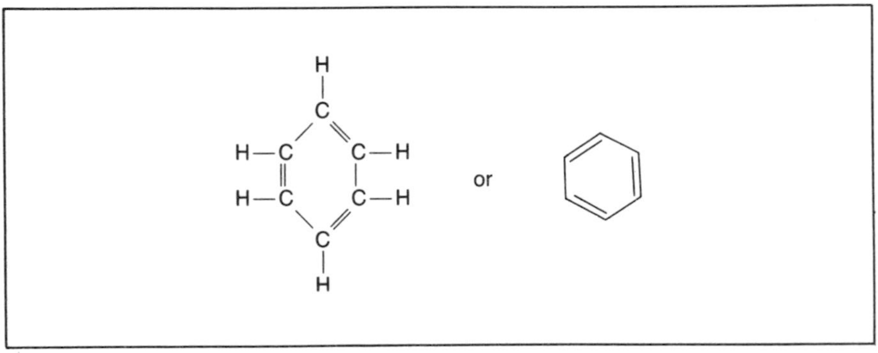

Fig. 1–13   Benzene, C_6H_6

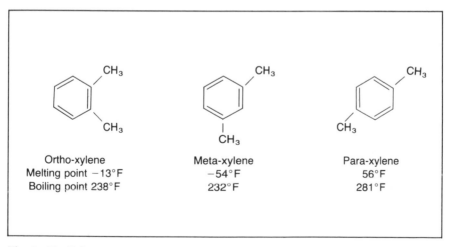

Fig. 1–14 Toluene, $C_6H_5CH_3$

Now go one step further. When two hydrogens are replaced on a benzene molecule (di-substitution), three isomers can occur. (Tri-substitution is more complicated still.) Consider xylene, for example, $C_6H_4(CH_3)_2$. That compound is a benzene ring, with methyl groups ($-CH_3$) replacing two hydrogens. As shown in Fig. 1–15, the replacement can be in one of the three (and only three) patterns.

Each of the isomers, called *ortho-*, *meta-*, and *para-xylene*, has unique physical properties. Two such properties are the melting points (when xylene turns from solid to liquid) and the boiling points (when xylene changes from liquid to gas). These characteristics figure importantly when the xylene isomers are separated (*see* Chapter III). *Mixed xylenes*, a commonly traded commodity, are the combination of the three isomers.

Ortho-xylene	Meta-xylene	Para-xylene
Melting point −13°F	−54°F	56°F
Boiling point 238°F	232°F	281°F

Fig. 1–15 Xylenes

Di-substituted benzene such as xylene can be compared to a litter of puppies. They are all dogs, but each one behaves differently. If the puppies were like monosubstituted benzenes (e.g., toluene), they would all be clones.

Heterocyclics

Draw an alicyclic or aromatic compound, which has a chain of carbon atoms in a closed ring. Then replace one or more of the carbon atoms with some other kind of atom (usually oxygen, nitrogen, or sulfur), and you have a *heterocyclic* compound. Ethylene oxide, CH_2OCH_2, is one of the simplest of the heterocyclic series, since it's a three-atom ring. (Anything smaller is not a ring.) As with alicyclics, five- and six-atom heterocyclic rings are more stable than the three-atom rings. Some of the more common heterocyclics are shown in Fig. 1–17.

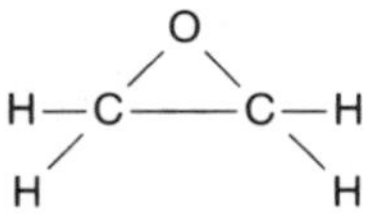

Fig. 1–16   Ethylene oxide, CH_2OCH_2

You must move one step from the road map of organic chemistry to complete this discussion. A large portion of petrochemistry is based on substituting oxygen for hydrogen in the Fig. 1–4 molecules. Although the resulting molecules do not exactly fit the definitions just covered, the family descendency from the various aliphatics and carbocyclics will be apparent. The simplest first step is to look at the alcohols.

Alcohols

The hydroxyl group, —OH, is the signature of the alcohol family. By removing a hydrogen and adding a hydroxyl group to an ethane molecule, you can form ethyl alcohol, C_2H_5OH. Methyl alcohol (sometimes called wood alcohol because wood at one time was its commercial source) is methane hooked up with the hydroxyl group, CH_3OH. You can go blind by drinking either of these alcohols, but the condition will be permanent with the latter. There's another kind of alcohol that can make you feel good. But you rub this one on the outside instead of ingesting it: isopropyl alcohol. This molecule is a good example of both alcohol and isomerism, as shown in Fig. 1–18.

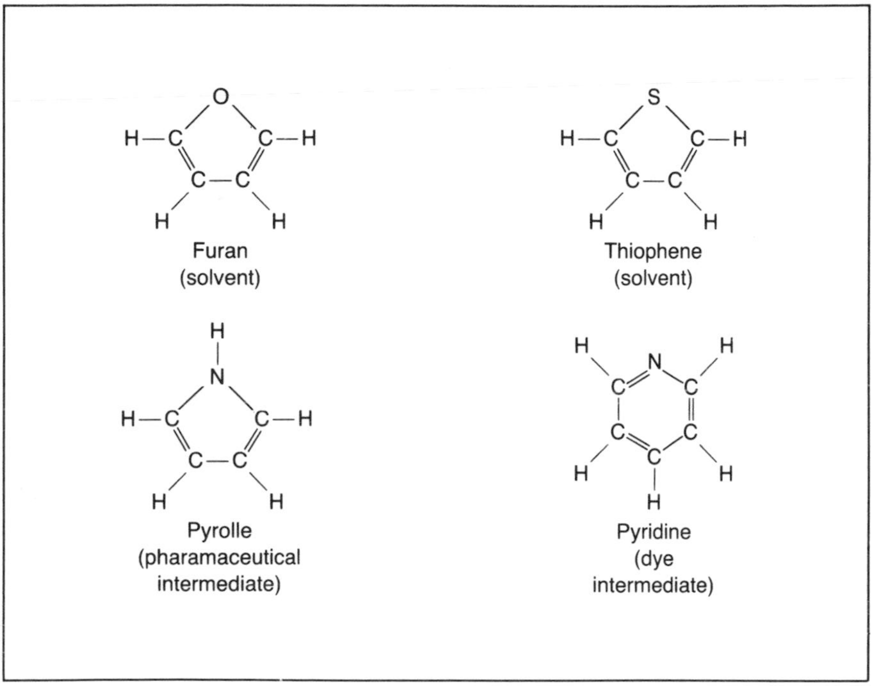

Fig. 1–17 Heterocyclics

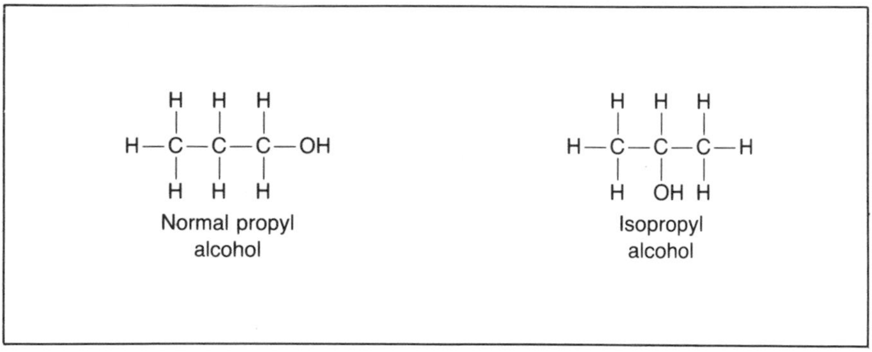

Fig. 1–18

For a very complicated reason (which you're better off not wanting to know), a carbon atom will attach itself to no more than one hydroxyl radical. However, each carbon atom in a molecule can have its own hydroxyl radical, as shown in Fig. 1–19.

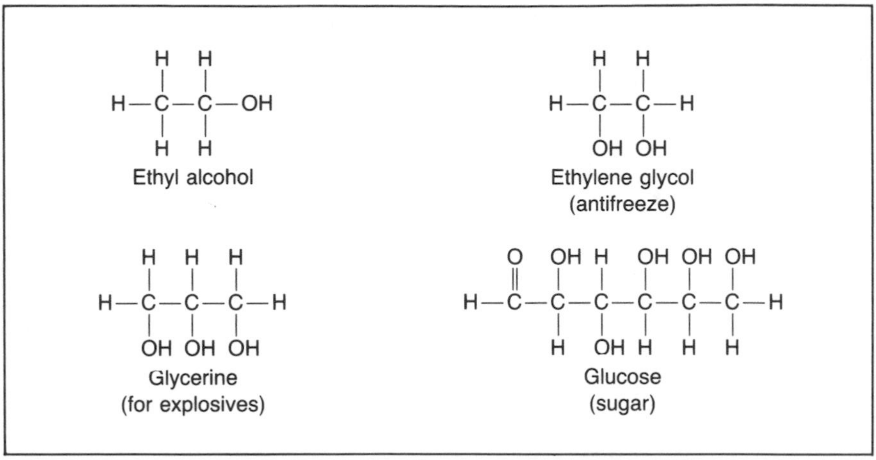

Fig. 1–19

Oxidation

The route to many petrochemicals is *oxidation*—the reaction of an atom or a molecule with oxygen. The oxygen can come from air or from another compound that readily gives up its oxygen (such as hydrogen peroxide, H_2O_2). If the reaction goes to the extreme (that is, complete oxidation of an organic compound), you always end up with oxygens connected to every carbon, forming carbon dioxide and water. Burning (combustion) is an example of complete oxidation. Your bodily functions also are a good example. For instance, if ethyl alcohol (vodka) is ingested, the ultimate result is a chemical imbalance in your system: $C_2H_5OH + 3O_2 \longrightarrow 2CO_2 + 3H_2O$. That's why you can relieve a hangover somewhat by breathing through an oxygen mask. It restores the normal oxygen balance to your body (and head).

In petrochemicals, partial oxidation is usually more desirable. Then one or more of the hydrogens springs loose to form H_2O; the empty bond is taken up by an oxygen. *Aldehydes*, for instance, have as their signature a grouping made of carbon, hydrogen, and oxygen in this configuration:

$$\overset{O}{\underset{}{\overset{\|}{-C}}}\!\!-H.$$ This grouping is always at the end of the carbon chain.

The *ketone* signature, $-\overset{O}{\overset{\|}{C}}-$, is never at the end of the chain. The oxygen atom with a double bond replaces two hydrogens. For contrast, a family of first and second cousins (a paraffin, an olefin, an alcohol, an aldehyde, and a ketone) is shown in Fig. 1–20. Note the signatures

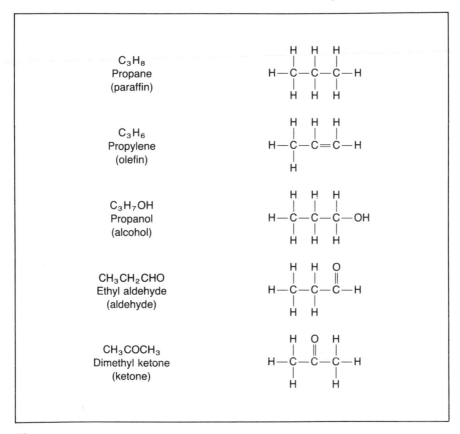

Fig. 1–20

—OH, —CHO, and —CO— imbedded in the chemical formula for propyl alcohol, ethyl aldehyde, and dimethyl ketone.

Compounds formed by adding the *carboxyl group*, $-\overset{\overset{\text{O}}{\|}}{\text{C}}-\text{OH}$, at the end of a chain are called *carboxylic* (car-box-ill-ick) acids. The carboxyl group signature is the dulcet-looking —COOH, but some of the compounds may not be so pleasant. Formic acid, HCOOH, is the main compound that ants inject in you when they bite; acetic acid, CH_3COOH, is the acrid smelling ingredient in vinegar. You can think of carboxylic acids as being third down the line on the route to oxidizing a paraffin completely:

Ethane	C_2H_6	
Ethyl alcohol	C_2H_5OH	Increasing
Acetaldehyde	CH_3CHO	oxidation
Acetic acid	CH_3COOH	state
Carbon dioxide and water	$CO_2 + H_2O$	↓

Other Compounds

The never-ending, ever-expanding lists of everything else should be the largest section of this chapter. It covers all of the compounds and classes for which students take the many advanced courses in chemistry. You will be relieved to know that in this section the discussion has been limited to six classes of compounds that are common to the petrochemicals business.

Anhydrides are derivatives of the carboxylic acids. Remember the grouping, $-\overset{\overset{\text{O}}{\|}}{C}-OH$? If you have two of them on adjacent carbon atoms and remove two hydrogens and an oxygen (enough to make water), you will form the anhydride grouping shown in Fig. 1–21.

Esters use that same carboxyl grouping, but they open up a more complex world by attaching any of a variety of organic groups in place of the hydrogen in the carboxyl signature, $-\overset{\overset{\text{O}}{\|}}{C}-OH$. This grouping makes it —COOR instead of —COOH. In Fig. 1–21, the notation "R" is used in the ester signature to imply another radical. (Using R is a common shorthand expedient of the chemist.) Two familiar esters are shown in Fig. 1–21: methyl acetate, a solvent for fast-drying automobile lacquers, and acetyl salicyclic acid, commonly called aspirin. (Not only will that relieve a headache physically but maybe figuratively too. If you study it closely, you may be able to identify a methyl group, an ester group, a benzene ring, and a carboxyl group.)

Ethers are simple. These chemicals are similar to water molecules (H—O—H) where the two hydrogens are replaced by two radicals (which may or may not be of the same type). There are also cyclic ethers. Ethylene oxide (in Fig. 1–16) is an oxygen connected to two CH_2 groups that are attached to each other. Ethylene oxide is important as a petrochemical *intermediate* (a compound made enroute to making one or more other compounds).

Diethyl ether is another example of an ether, the name being apparent from the connection of two ethyl groups to the ether signature,

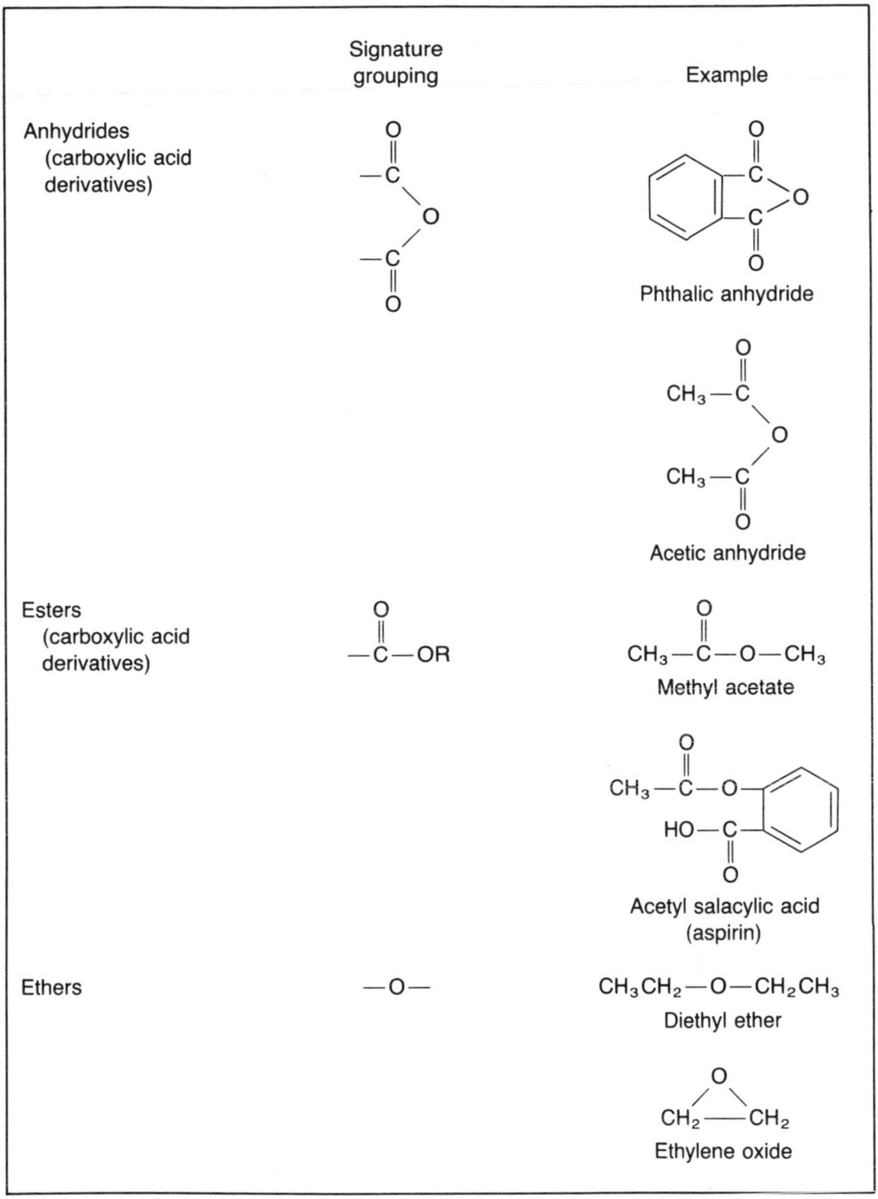

	Signature grouping	Example
Anhydrides (carboxylic acid derivatives)		Phthalic anhydride Acetic anhydride
Esters (carboxylic acid derivatives)		$CH_3-C-O-CH_3$ Methyl acetate Acetyl salacylic acid (aspirin)
Ethers	—O—	$CH_3CH_2-O-CH_2CH_3$ Diethyl ether Ethylene oxide

Fig. 1–21

—O—. Diethyl ether is the doctor's ether in its gaseous form. This ether is very volatile and is extremely flammable. In fact, precautions

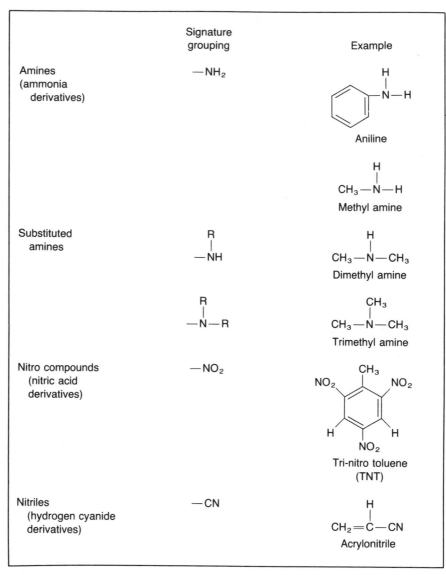

Fig. 1–22   Nitrogen-based chemicals

must be taken in an operating room when ether is administered. Any spark, even from static electricity, can cause an explosive reaction of the ether with the oxygen in the air (a gruesome thought if you're the patient with a lung full of diethyl ether).

There is a large family of organic compounds just outside of petro-chemicals that revolves around nitrogen. Sometimes these compounds even intermarry. For example, the *amines* shown in Fig. 1–22 are commonly found in the petrochemical industry. *Aniline* ($C_6H_5NH_2$), an important dye intermediate, is a typle example. Amines are generally formed by reactions involving *ammonia* (NH_3) or ammonium hydroxide (NH_3 bubbled into water). Similarly, the *nitro* compounds are charac-terized by the signature grouping of nitrogen and oxygen, —NO_2. Usu-ally the NO_2 comes from nitric acid, like the reaction between nitric acid and toluene to make trinitrotoluene (TNT). Finally, a small group of near-petrochemicals are the *nitriles*. These are compounds with the signature group —CN. The family success story here is acrylonitrile, a compound used extensively to manufacture tires, plastics, and synthetic fibers.

That's it! This is, by far, the toughest chapter in the book. If you made it this far, the rest will be easy. There will be more chemistry as you go through each chapter, but the doses will be so small they'll go down in a quick swallow. So move on!

II.

BENZENE

Oh my how many torments lie in the circle of a ring

"The Double Gallant,"
Colley Cibber (1671–1757)

Why start out with *benzene*? The apparent answer is that benzene is one of the basic building blocks in the petrochemicals industry, along with ethylene, propylene, and a few others. To complement that, benzene, more than any of these other chemicals, comes from a variety of sources (steel mill coking, petroleum refining, and olefin plants).

Historical Perspective

Benzene was first isolated and identified in 1825 by Michael Faraday during his scientific heyday at the Royal Institute in London. However, this chemical proved to be an enigma to chemists for more than a century. The valence rules of carbon and hydrogen demand that benzene have the characteristic alternating single and double bonds in the carbon ring. However, the benzene molecule does not behave in the precise way that other molecules with double bonds do.

In chemical reactions, the carbon-carbon bonds in the benzene ring act like single bonds. In terms of bond energies, they respond like an average between single- and double-bonded carbons. Considering these characteristics, the German chemist August Kekulé offered a very appealing theory. In 1865, he suggested that single and double bonds continuously trade places with each other—they oscillate or resonate. In 1931, the famous Linus Pauling helped support this theory, using quantum mechanics. There are still some loose ends, but no good alternative theory has been developed.

Benzene had limited commercial value during the nineteenth century and was used primarily as a solvent. In the twentieth century, gasoline blenders found that benzene had good octane characteristics. As a consequence, there was a large economic incentive to recover the benzene produced as a byproduct of the coke ovens in the steel industry. During World War II, other chemical uses for benzene emerged such as making explosives. Initially the coke-oven benzene was diverted from gasoline blending to the chemical companies. By mid-century, the refining industry was recovering huge quantities of the benzene in gasoline to keep up with chemical demands. There's some irony in the fact that the largest consumer of benzene, the petroleum refining business, ultimately turned out to be its largest supplier.

The increasing demands for benzene for petrochemicals led to two developments. Catalytic reforming and benzene recovery in refineries were improved, and *toluene dealkylation*, a technique for converting toluene to benzene, was commercialized. The latter process goes in and out of vogue as the economic winds blow to and fro. Another fortuitous contribution to the supply began in the early 1970s, when benzene byproducts were manufactured from olefin plants, using heavy gas oil as feedstock.

Benzene from Coal

An important raw material for manufacturing steel is coke, a nearly pure carbon compound. To supply coke for the industry, steelmakers developed the process of destructive coal distillation.

The chemical makeup of coal is predominantly a mixture of very *high molecular-weight, polynuclear-aromatic* compounds. That mouthful is a common expression used to describe heavy hydrocarbon compounds. High molecular weight refers to the number of atoms attached together in the molecule (in this case, carbon and hydrogen). Ethane, C_2H_6, is low molecular weight; $C_{30}H_{30}$ is high. The term polynuclear aromatic refers to the preponderance of C_6-type rings in the molecule, as shown in Fig. 2–1.

Since these molecules are large and have the multiple-ring feature, the ratio of carbon to hydrogen is high compared to the other hydrocarbons encountered up to this point. In ethane the ratio is 1:3; the compound in Fig. 2–1 is almost 2:1. In the destructive distillation process, coal is heated to 2,300–2,700°F in the absence of air. At those temperatures, the large molecules begin to crack, forming smaller organic compounds (many of which are liquids or gases at room temperature) and pure carbon (coke for the steel furnaces).

Because of the high carbon/hydrogen ratio, one ton of coal yields about 1,500 pounds of coke and about 500 pounds of coal gas, oil, and tar. Prior to electricity, coal gas was the primary source of municipal lighting. Gas lights lined the streets of the great cities in 1900. Coal tar is a solid at room temperature and is often used as a roofing material or for paving roads.

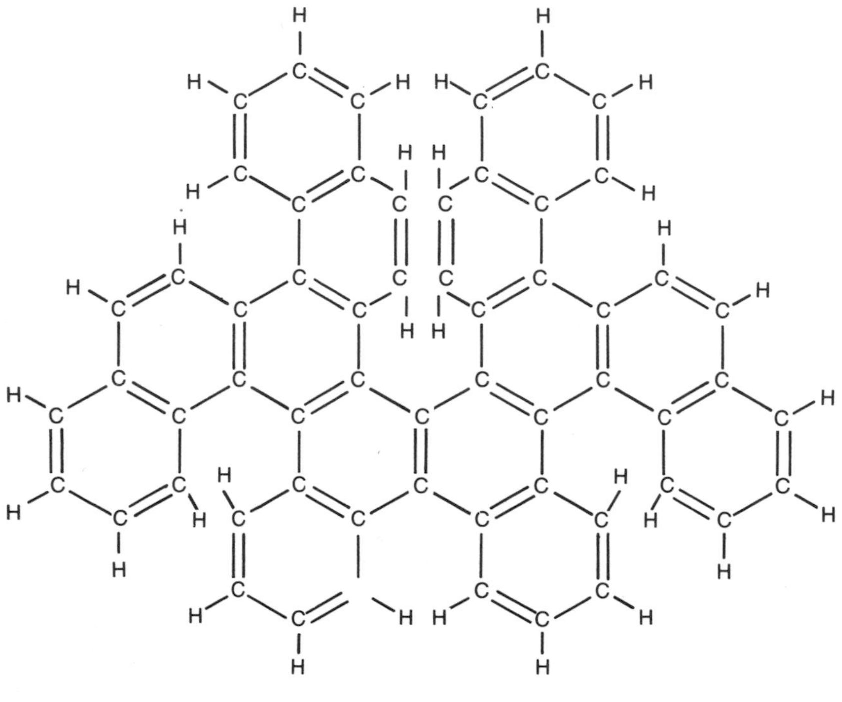

$C_{57}H_{32}$

Fig. 2–1   A polynuclear aromatic

The coal oil byproduct is predominantly a mixture of benzene (63%), toluene (14%), and xylenes (7%), resulting directly from the basic benzene ring that remains intact during the cracking process. Therefore, steel companies have become important suppliers of *BTX* (benzene, toluene, and xylenes) by exploiting a valuable byproduct.

Benzene in Petroleum Refining

Benzene comes from two sources in the typical petroleum refinery. First it can be found in the crude oil pumped from the ground. However, a typical crude has only 0.5–1% pure benzene. Generally that is not enough to justify the equipment needed to extract it.

The second, more commercial source of benzene is the *catalytic reforming* process. The primary purpose of this process is to make quality gasoline components out of low-octane naphtha. To be more specific, the feed to a catalytic reformer is a mixture of paraffins and cyclic compounds in the C_6–C_9 range, generally called naphtha. Typically, a catalytic (cat) reformer changes the naphtha as shown:

	Volume %	
	Feed	Product
Paraffins	50	35
Naphthenes	40	10
Aromatics	10	55

The cat reforming process changes the chemical composition of the naphtha by causing the following reactions (illustrated in Fig. 2–2): paraffins are converted to isoparaffins and to naphthenes; naphthenes are changed to aromatics (including benzene). These reactions are favorable because the end products (isoparaffins, naphthenes, and aromatics) have higher octane numbers than the compounds from which they are created.

Other changes are not so good. Paraffins and naphthenes can crack and form butane and lighter gases. Also, some of the attached radicals can break from the naphthenes and aromatics and make butanes and lighter gases. The butanes and lighter gas usually have a much lower economic value than gasoline.

Process Descriptions.   These reactions are promoted and controlled by pumping the naphtha feed through reactors under high pressures (200–800 psi*) and high temperatures (850–950°F). However, those conditions do not act alone; they work in the presence of a catalyst. Each spherical reactor (shown in Fig. 2–3) is packed with pellets made of aluminum or silica that are coated with platinum. Various reactions can occur when the naphtha contacts the platinum, depending on the temperature and pressure in that particular reactor. Generally there are several reactors, so different operating conditions can be used to promote the desired reactions.

From the diagrams in Fig. 2–2 you can see that some reactions give off hydrogen and others use hydrogen. Therefore, when the product leaves the last reactor, the hydrogen is separated from the remaining product. Then the hydrogen is recycled and is mixed with incoming feed. This process provides an abundant supply during the reactions. The hydrogen abundance also prevents coke formation, which results from high-temperature naphtha pyrolysis (cracking due to exposure to high temperatures). If coke is deposited on the catalyst, the catalyst is

---

*Pounds per square inch.

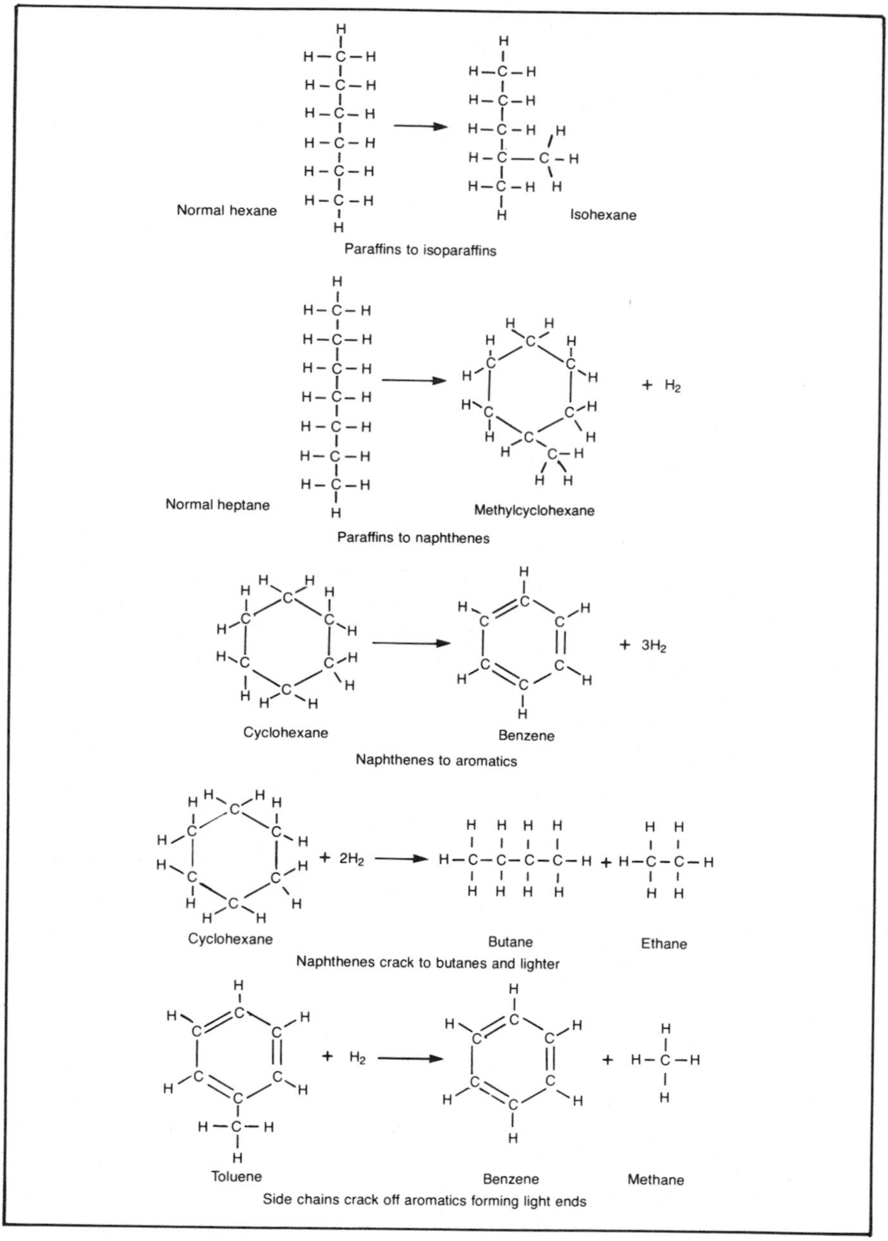

Fig. 2–2 Cat reforming reactions

quickly deactivated. However, with excess hydrogen, coke unites with hydrogen to form light paraffins such as methane and ethane.

Eventually, the catalyst does become deactivated. Then the reactor must be shut down and regenerated. Otherwise, the amount of feed converted to the desired products declines rapidly. Regeneration is done primarily by pumping very hot air through the reactor. The oxygen in the air reacts with the carbon deposited on the catalyst (coke) and forms carbon dioxide, which is just blown into the atmosphere. After a lot of deactivation and regeneration, the catalyst starts to collapse and must be replaced.

The amount of benzene produced in a reformer depends on the composition of the feed. Every crude oil has naphtha with a different *PNA* (paraffin, naphthene, and aromatics) content. In commercial trading, the PNA content is often an important quality specification because it determines to a large extent the qualities of the naphtha. High aromatic content makes a good gasoline blending component. High naphthene content makes good reformer feed.

The amount of benzene is also determined by the mode in which the reformer is run. For example, setting the operating conditions to maximize benzene production generally increases the production of *light ends*—butanes and lighter gases. That's okay if you're not concerned about losing the other components used in gasoline. Therefore, the yields from a reformer are a function of the feed composition, the operating conditions, and the economic incentives to produce the various components of the products (the benzene, gasoline components, and light ends).

Downstream of the reactors and the hydrogen separator, the product is fed to a *prefractionator* or *stabilizer* where the butanes and lighter gases are removed. In many refineries the bottom product of the fractionator, called *reformate,* is used as a gasoline component with the remaining benzene. Benzene is a very desirable component because it has a high octane number and improves the performance characteristics of finished motor gasoline.

In refineries that recover benzene as a separate product, the reformate must be processed further in what is variously called a *BEU* (benzene extraction unit) or an *ARU* (aromatics recovery unit). As shown in Fig. 2–4, the reformate is charged into a feed preparation unit that separates the benzene from the hydrocarbons that have similar boiling temperatures. Theoretically, distilling units could fractionate a nearly pure benzene stream, but they would be inordinately expensive to build and operate. Instead, the benzene is concentrated by the feed prep columns to a narrow boiling-range mixture of compounds called *benzene concentrate.* Benzene concentrate typically is 50% benzene, plus some C_5, C_6, and C_7 compounds that boil at about 176°F (the boiling tem-

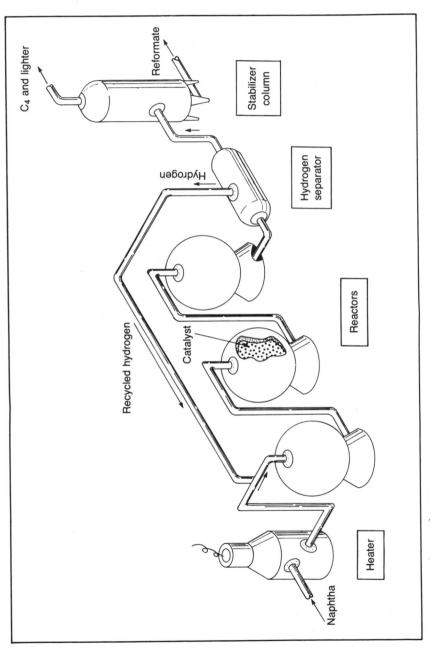

Fig. 2–3 Catalytic reformer

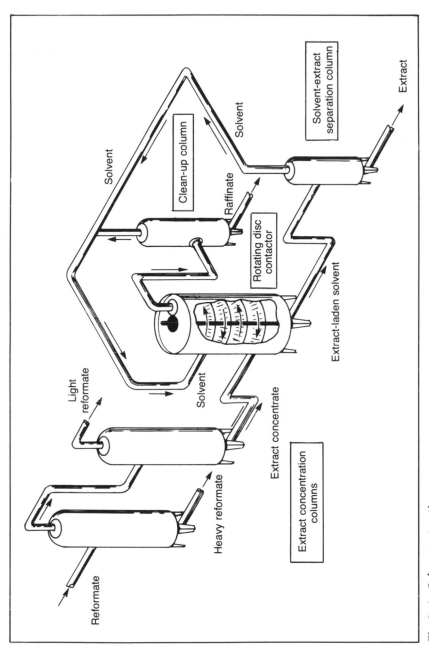

Fig. 2–4 Solvent extraction

perature of benzene). This stream is then fed to a *solvent extraction unit* like that shown in Fig. 2–4.

Solvent Extraction. Certain compounds have the remarkable characteristic of selectively dissolving some compounds and ignoring others. A familiar example is to drop a spoonful of table salt into half a glass of paint thinner. The salt sinks to the bottom in a pile. Mix it or shake it; it still settles to the bottom in a pile. The salt cannot be separated from the paint thinner completely. Now add half a glass of water and stir. The salt disappears because it dissolves in the water. Then to separate the salt water, you can carefully pour off the paint thinner, which is floating on top of the water. If you let the water evaporate for a few days, almost pure salt will remain in the bottom of the glass. In petrochemical language, water is the solvent in this example. Paint thinner with salt in it is the feed, and salt is the extract.

Solvent extraction of benzene works the same way, but the common solvents are sulfolane, liquid sulfur dioxide, and diethylene glycol. The paint thinner/salt/water process described above is called a *batch* solvent process, since it consists of sequential steps that can be reiterated batch after batch. Some commercial solvent extraction processes still operate in this manner. A batch of benzene concentrate is mixed with a solvent. The benzene dissolves in the solvent; the solvent separates from the remaining hydrocarbons. Then the benzene-laden solvent is drawn off and is distilled (fractionated). The last step is made easier by using a solvent with a boiling point much different from the benzene.

Knowing how the batch method works, you'll find the continuous-flow process just as simple. In Fig. 2–4, the benzene concentrate is pumped into the bottom of a vessel or column with a labyrinth of mixers inside. Sometimes the mixers are mechanically moved, such as in a rotating disc contactor. The solvent is pumped into the top, and most of it works its way to the bottom of the vessel. The benzene concentrate moves toward the top. As the two slosh past each other, the benzene is extracted from the concentrate and is dissolved in the solvent (which ends up in the bottom of the vessel).

In the continuous-flow process, the benzene-laden solvent is handled just like in the batch method. It is fractionated to separate the benzene from the solvent, providing commercial-grade benzene. Then the solvent is recycled to the mixing vessel. The remnant hydrocarbons from the top of the mixing vessel are often called *benzene raffinate*, a misleading name. Benzene raffinate contains no benzene. It is the leftover product after the benzene is removed, and it usually ends up as a gasoline blending component.

Benzene from Olefin Plants

Ethylene and propylene can be manufactured by cracking naphtha or gas oil in an olefin plant.* One of the byproducts of cracking these feedstocks is benzene.

Naphthas and gas oils consist of molecules with carbon counts of five, ten, twenty, or more. Olefins are created by heating the molecules to a temperature where they crack, forming (among other things) the desired C_2H_4 or C_3H_6. The larger carbon count molecules, C_{10} and higher, usually contain multiple benzene-ring structures, not unlike the coal configuration previously described. When the molecules break up, the benzene rings can be freed intact, creating benzene and other aromatics. (The process is not unlike destructive distillation of coal.) The benzene leaves the olefin plant mixed with other gasoline components, and generally the stream is sent to the benzene extraction unit for recovery.

Benzene from Toluene Hydrodealkylation

Since toluene is nothing more than benzene with a methyl radical, creating one from another should be relatively easy and indeed it is. Benzene, toluene, and xylenes are coproduced in coke-making, reforming, and olefin-plant operations. The ratio of production is rarely equal to the chemical feedstock demands. Usually the ratio of benzene to the other two is too low. One method for balancing supply and demand is *toluene hydrodealkylation*.

The word hydrodealkylation is less ominous than it appears. Alkane is a synonym for paraffin. Alkylation is the process of adding a paraffin radical (like the methyl or ethyl radical) to another compound. Dealkylation is removing it. Hydro indicates the replacement is hydrogen.

In the toluene hydrodealkylation process shown in Fig. 2–5, toluene is mixed with a hydrogen stream. This stream is heated and is pumped into a reaction vessel. The vessel, like a cat reformer reactor, is packed with a platinum catalyst (platinum or platinum oxide on a silica or aluminum-oxide base). At the operating temperatures (930–1,100°F) and pressures (600–900 psi), the methyl radical pops off as the toluene passes over the catalyst. Then hydrogen fills out the benzene valence requirements.

Products leaving the reactor pass through a separator, where unreacted hydrogen is removed and recycled to the feed. Further fractionating and treating separates the benzene from the methane and any other

---

*See Chapter V for a detailed explanation of these production methods.

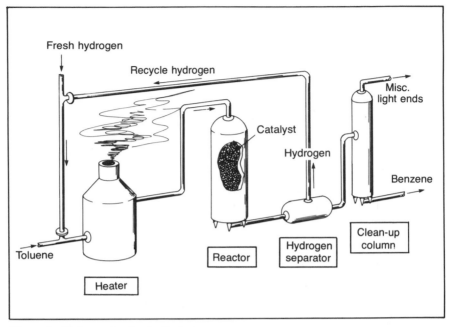

Fig. 2–5 Toluene hydrodealkylation

hydrocarbons in the mix. The yield of benzene from toluene in a hydro-dealkylation plant is typically 96–98%.

Handling Benzene

Benzene is a clear, colorless, flammable liquid with a characteristic sweet odor. It burns with a smokey flame, as do other hydrocarbons with high aromatic content. (That's why kerosenes with high aromatic content do not make good jet fuel or burning grade kerosene; they create too much smoke.) Benzene is only slightly soluble in water, and it boils at 176°F.

The two commercially traded grades of benzene are crude and industrial pure. Crude benzene has some miscellaneous hydrocarbons mixed in with it, in the range of ±7°F of the benzene boiling temperature. Industrial-pure benzene impurities are confined to about 0.5%; the boiling range is ±3.6°F. The benzene also is treated in clay to remove most miscellaneous organic compounds, particularly thiophene.

Benzene is shipped by tank cars or trucks, barges, tankers, and smaller containers. Transfers from one vessel to another are in a closed system because benzene is a poisonous substance with acute toxic effects. It can kill you in 5–10 minutes, if you breathe too much.

Use Patterns

The three major uses of benzene are the manufacturing processes for styrene, cyclohexane, and cumene. Polymers and all sorts of plastics are produced from styrene; cyclohexane is the next step to synthetic fibers of many kinds. Cumene leads to phenol, which ultimately ends up in resins and construction adhesives such as plywood. Less significant volumes of benzene are used in the processes for maleic anhydride, nitrobenzene and dodecylbenzene.

Benzene technology (making and recovering it) and handling are mature operations. Presently, little development is going on, however, derivative applications are still being studied. The same is true for toluene and xylene, which are covered in the next chapter.

III.

TOLUENE AND THE XYLENES

Into fire, into ice.

Divine Comedy,
Dante (1265–1321)

The chemistry and hardware involved in making toluene and xylenes are for the most part the same as their sibling, benzene. While that may be true, there are a few chemical principles that can be demonstrated better using toluene. But more importantly, the separation processes for purifying toluene and xylenes are different.

Toluene

Like benzene, toluene is manufactured by cat reforming processes, coke production, and olefin-plant operations and by recovering the small amount found naturally in crude oil. In the last three, toluene is clearly a byproduct, just like benzene. However, most toluene comes from cat reforming, a process controlled by several variables: the feed composition and the operating conditions in the reactors.

Some compounds are more suitable for reforming into toluene. These precursors include cyclohexane, methyl cyclohexane, ethyl cyclopentane, and dimethyl cyclopentane. Fig. 3–1 shows the three basic reactions that occur on the way to toluene: ring opening, dehydrogenation, and cyclization. Ethylcyclopentane and methylcyclohexane are usually the most abundant precursors. A reformer feed that is naturally rich in these compounds will have a high toluene yield.

Unfortunately (for the toluene merchant), there is not much attention paid to optimizating toluene production in refineries, for several reasons:

1. More toluene usually means less benzene.
2. The choice of the naphtha feed depends on the crude oil selected to distill. The type of crude determined by factors other than reformer operations because the reformer feedstock fraction is only a small part of crude oil.
3. Most toluene and xylene end up as gasoline blending components.

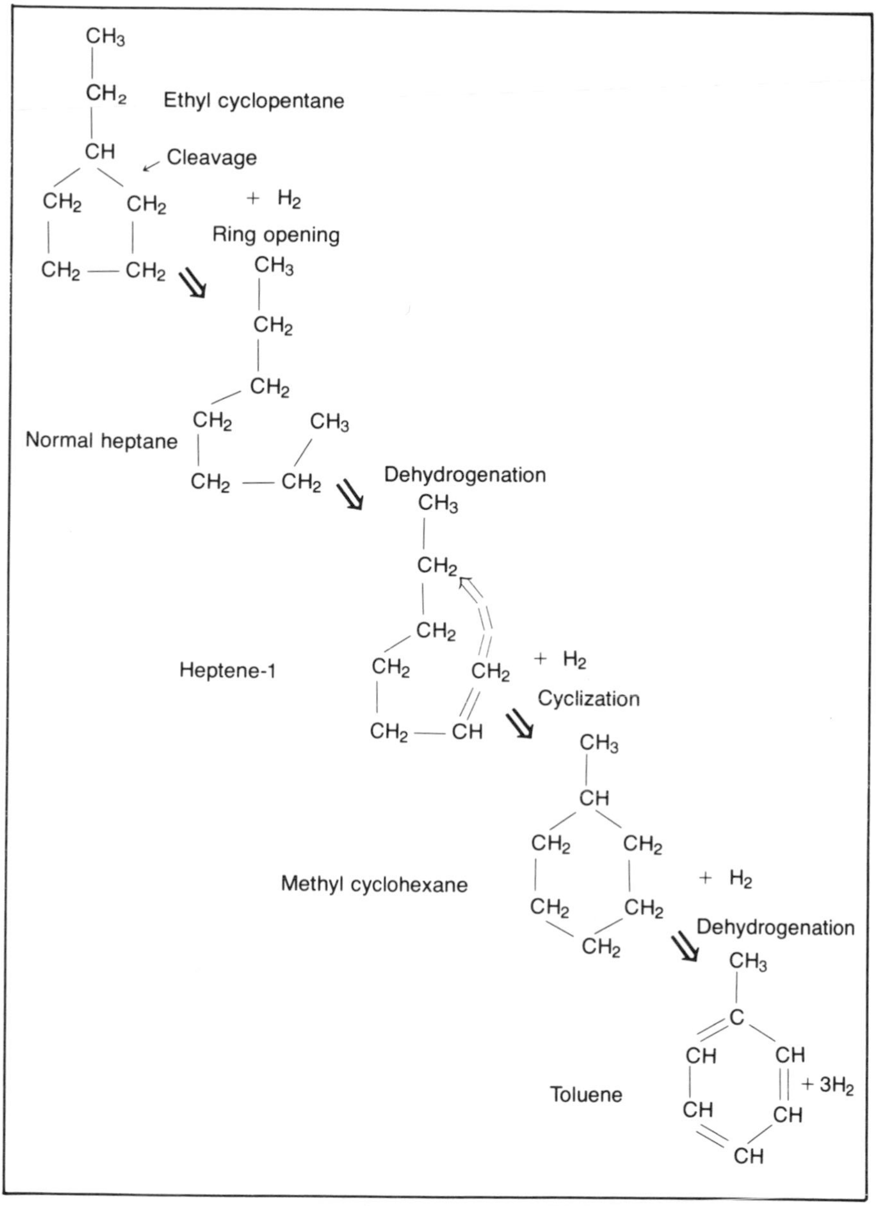

Fig. 3–1 Toluene production by catalytic reforming of ethyl cyclopentane

In any case, the reformate stream leaving the cat reformer can have ample amounts of toluene. Toluene can be separated from the other components by solvent extraction with the benzene or in individual streams. The boiling points of benzene and toluene are far enough apart that the feed to the solvent extraction can be split rather easily into benzene and toluene concentrates. Alternatively an aromatics concentrate stream can be fed to the solvent extraction unit, and the aromatics produced can be split into benzene and toluene streams by fractionation. Both schemes are popular.

Azeotropic Distillation

Another process, called *azeotropic distillation*, can be used to split toluene from the other hydrocarbons that have boiling points near toluene. This distillation process can be much more efficient than solvent extraction when the toluene concentration is high.

In azeotropic distillation, the liquid raffinate (everything in the toluene concentrate but the toluene) is removed by a coliquid. The coliquid causes the raffinate to vaporize at a lower temperature, making it easier to distill. An example may help explain the process. An occasional automotive problem is water in the gasoline, usually caused by warm moist air in a nearly empty gas tank. This water condenses when the weather turns cold. Water in gasoline makes starting the engine difficult and causes sputtering because the water does not vaporize in the carburetor. An over-the-counter remedy is dry gas, which is nothing more than ethyl alcohol. Water dissolves in the alcohol. Together these chemicals act like gasoline as they go through the carburetor because both substances vaporize at a lower temperature than either water or ethyl alcohol alone. In this analogy, the coliquid is ethyl alcohol, the (toluene) extract is gasoline, and the raffinate is water.

The coliquid in azeotropic distillation increases the volatility of all of the chemicals except the extract, making separation in a distillation column easy. When azeotropic distillation is used for toluene, the coliquid is usually a mixture of methyl ethyl ketone (MEK) and water (10%). The coliquid and the toluene concentrate are mixed and heated, then they are charged to a distillation column (Fig. 3–2). The liquid raffinate, which consists of paraffins and naphthenes, dissolves in the MEK/water and vaporizes about 20°F lower than normal. The vapors work up the distilling column, and the toluene moves down as a liquid. This process occurs despite the fact that the paraffins and naphthenes have nearly the same boiling temperatures as toluene because the coliquid causes the reaction. Toluene from the bottom of the distillation column is treated for residual MEK/water or other contaminants, as shown in Fig. 3–2. Further fractionation for purity is optional.

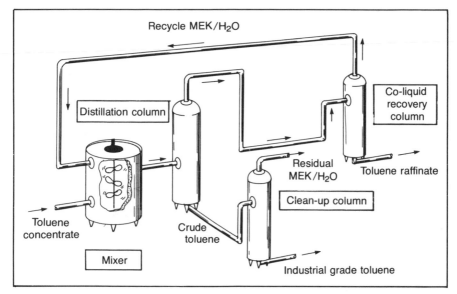

Fig. 3–2   Azeotropic distillation of toluene

Commercial Use

There are three grades of toluene commercially traded. Industrial grade is 95–98% toluene, boiling within one degree of the toluene boiling point, 213°F. Nitration grade is 99% pure, and it boils within a half degree of 213°F. (The term "nitration" is a hangover from the specification developed during World War II for manufacturing trinitrotoluene, or TNT.) Research grade is 99.9% pure toluene.

Use Patterns of Toluene

During World War II, interest grew most rapidly in two toluene applications that are militarily related, but not chemically. Since toluene has high octane characteristics, manufacturing and recovering it for aviation gasoline blending was crucial. At the same time, the demand for TNT explosives soared. Ironically, the chemistry that develops good octane characteristics has nothing to do with the chemistry of explosives.

In later years, toluene became an important source of benzene through the toluene hydrodealkylation unit. Toluene as a gasoline and as an aviation-gasoline blending component continues. There are myriad applications for toluene as an industrial solvent. It is also an ingredient in manufacturing toluene *diisocyanate*, the precursor to polyurethane foams. Other derivatives include phenol, benzyl alcohol, and benzoic acid.

Xylene

You might think that to complete the aromatics family, the X in BTX would be the triplets, ortho-, para-, and meta-xylene. However, there is another member called ethylbenzene. Ethylbenzene has the xylene carbon and hydrogen count, C_8H_{10}. However, it's a benzene ring with an ethyl group ($—C_2H_5$) attached, rather than two methyl groups ($—CH_3$).

The manufacturing process for xylenes is similar to benzene and toluene: cat reforming, olefin plants, coke production, and a small amount naturally resident in crude oil. However, the unique aspect of xylenes is the techniques that separate them from each other.

Process Description. In a refinery where all aromatics are recovered (i.e., not dumped into gasoline), the processing scheme is generally similar to that in Fig. 3–3. The reformate stream coming from the cat reformer contains the aromatics that have been created, together with miscellaneous naphthenes (alicyclics) and paraffins (aliphatics). BTX are recovered as a mixture in the solvent extraction unit. The boiling points of benzene, toluene, ethylbenzene and ortho-xylene are far apart, so they can be separated by simple fractional distillation. However, meta- and para-xylenes must be removed by other methods.

	Boiling temperatures, °F
Benzene	176.2
Toluene	231.4
Ortho-xylene	292.0
Meta-xylene	282.4
Para-xylene	281.0
Ethylbenzene	277.1

Ortho-xylene can be separated from the others by distillation. Ethylbenzene is only 3.9°F from para-xylene, but using superdistillation, tall, multitrayed columns (i.e., 200 feet and 300 trays), these substances also can be separated. However, with the 1.4°F spread between meta- and para-xylene, very expensive columns would be required. So alternative separation techniques have been developed: cryogenic crystallization and molecular sieves.

Cryogenic Crystallization. Although the boiling temperatures of meta- and para-xylene are close, their melting points (i.e., the temperature at which the liquids turn to crystals) are not. Meta-xylene crystallizes at -54.2°F and para-xylene at $+55.9$°F.

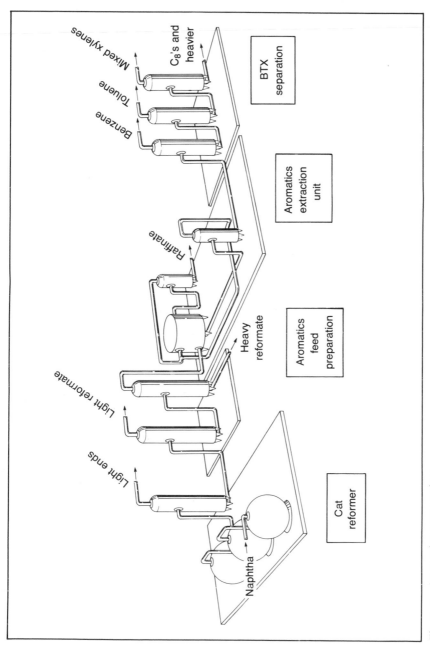

Fig. 3–3 Aromatics processing

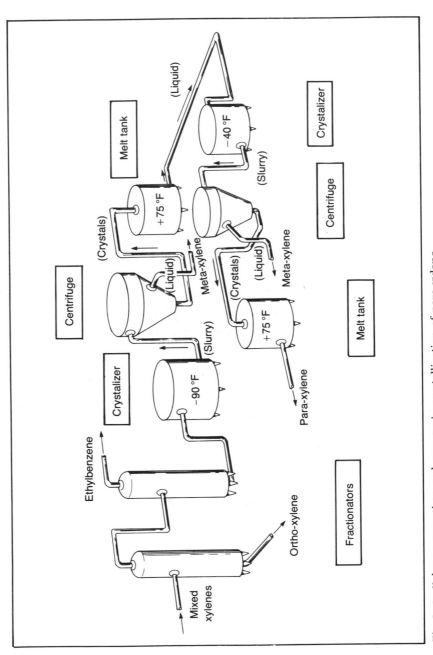

Fig. 3–4 Xylene separation and cryogenic crystallization of para-xylene

In Fig. 3–4, the processing scheme shows how ortho-xylene and ethylbenzene are split in fractionators. The mixed para- and meta-xylenes are then processed in a fashion similar to making good pot roast gravy. In order to keep grease out of the gravy, the beef drippings could be placed in the refrigerator for an hour or two. The grease solidifies and floats to the top; then it can be spooned off and discarded. Similarly, the mixed para- and meta-xylenes are cooled to about −90°F in a holding tank. At that temperature, para-xylene crystals form, creating a liquid-solid mixture like slush. When the crystals have grown sufficiently, the slush is put in a centrifuge. The spinning action permits the para-xylene to separate from the mother liquor, so-called because the crystals come out of the liquid.

To purify the para-xylene further (which still has some meta-xylene in it), the crystals are melted, cooled to about −40°F, and crystallized again. This time centrifuging results in about 95% pure para-xylene. The meta-xylene combined from both centrifuges is about 85%.

Molecular Sieves. Another commercial process has been developed using a material called molecular sieve, which can separate the para- and meta-xylenes. Molecular sieves are marble-sized pellets that have millions of pores. Para-xylene can fit in these spaces, but ortho-xylene cannot. The pore sizes are so small, they are measured in angstroms, which are 1×10^{-8} centimeters (0.00000001 cm). Molecular sieves of varying pore sizes are used in many other applications as well.

In a para-xylene plant, a column or bed packed with molecular-sieve pellets is used. The mixed para- and meta-xylenes are pumped through the bed. At first the stream coming from the bed is very low in para-xylenes, which are collecting in the sieves. Gradually the concentration of para-xylene in the effluent starts to build as the pores fill up. At some point, the operation is shut down. Then a fluid is pumped back through the bed to flush out the para-xylene. The process of pulling the para-xylene from the sieve is called *desorbing* or *desorption.* The flush fluid is called a *desorbent.* The desorbent used is chosen so that it can be easily separated from the para-xylene in a distilling column.

After desorption, the bed is heated to vaporize and remove the desorbent and the remaining para-xylene. Then the cycle can begin again. Para-xylene purity from this technique is about 99.5%.

Commercial Use. The composition of mixed xylenes depends on the cat reformer operations at a refinery and the type of crude oil and naphtha being processed. Typical composition is as follows:

Ethylbenzene	10–15%
Para-xylene	17–20%
Meta-xylene	46–52%
Ortho-xylene	18–24%

Mixed xylenes are commercially available in nitration grades with tolerances of 3 and 10°F, depending on the amount of hydrocarbon present. Purities of the ortho-, meta-, and para-xylenes are often a matter of negotiation between buyer and seller. Ethylbenzene as a separate stream is generally not a commercial commodity. Its primary use in petrochemicals is as an intermediate process stream for making styrene from benzene and ethylene.

The primary use of mixed xylenes is as an octane improver in gasoline. They also can be commercial solvents, particularly for industrial cleaning operations. Para-xylene is principally used to manufacture terephthalic acid and dimethyl terephthalate enroute to polyester fiber. Phthalic anhydride is made from ortho-xylene and in turn is used to manufacture polyester, alkyd resins and PVC plasticizers. Meta-xylene is sometimes converted to isophthalic acid, a chemical in thermally stable polyester and alkyd resins. There has been a greater supply of meta-xylene than is needed for its limited derivatives, so some is converted in isomerization units to ortho- and para-xylenes.

Xylenes are shipped by the same methods as benzene and toluene: tank cars and trucks, barges, and tankers. Pipeline movements are limited, but only by the small volumes going between any two points. Toxicological problems dictate handling in closed systems, as with benzene and toluene.

IV.

CYCLOHEXANE

A hen is an egg's way of making another egg.

Life and Habit,
Samuel Butler (1835–1902)

The petrochemical business is funny. Some companies use cyclohexane to make benzene and some use benzene to make cyclohexane. This chapter covers the latter.*

Interest in cyclohexane as a petrochemical blossomed in the Forties, after the processes for manufacturing caprolactam and nylon were developed by duPont in 1938. Very little cyclohexane occurs naturally in crude oil, certainly less than that required to satisfy caprolactam needs. Benzene, which has the right shape but not enough hydrogens, provided the solution to this shortage. Cyclohexane plants are not much more than vessels in which benzene molecules are hydrogenated with the help of a catalyst.

You'll recall that benzene has alternating double bonds. Hydrogen added to one of the carbons cascades around the benzene ring, so all of the carbons pick up a hydrogen. However, pressure and temperature are not effective methods of hydrogenation. Fortunately, several metals can promote this chemical reaction: platinum, palladium, nickel and chromium. The first two are highly active and can cause hydrogenation to occur at room temperature with only 15–20 psi pressure. Unfortunately, platinum and palladium are expensive metals. Most commercial processes use nickel or chromium. These catalysts are cheaper but require much higher temperatures and pressures, which are more expensive in terms of energy costs.

Sulfur can be a killer (literally) with hydrogenation catalysts. It will poison the catalyst, making it completely ineffective. Sulfur can show up in the feed (benzene) or in the hydrogen stream. The alternatives to protect the catalyst are to treat the feed or the hydrogen or to use a sulfur-resistant catalyst metal such as tin, titanium, or molybdenum.

---

*The former, cyclohexane to benzene, takes place in a cat reformer (*see* Fig. 2–1).

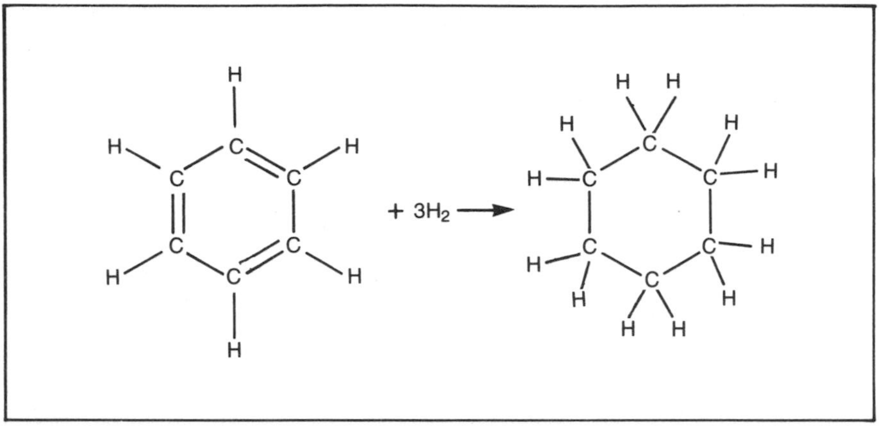

Fig. 4–1 Benzene hydrogenation to cyclohexane

The economic tradeoffs are additional processing facilities and operating costs versus catalyst expense, activity, and replacement frequency.

Process Description

The hardware use for hydrogenating benzene is shown in Fig. 4–2. The basic parts are three or four reactors in a series plus a separation section at the end. The reactors are vessels filled with a catalyst, one of the above metals deposited on charcoal or alumina pellets. The catalyst is packed loosely, so the feed can flow through from top to bottom.

The continuous flow process shown in Fig. 4–2 has a mixture of benzene, cyclohexane, and hydrogen that is heated to about 400°F, pressured to 400 psi, and pumped through the first reactor. The proportions of each feed depend on the type of catalyst used. On a once through basis, about 95% of the benzene is converted to cyclohexane.

Most hydrogenation reactions (including this one) are *exothermic*, i.e., they give off heat. To minimize byproducts, strict temperature control must be maintained. As the feed passes through the reactor, the temperature increases about 50°F. Therefore, the reactor effluent is cooled to 400°F in a heat exchanger. For the second pass more benzene is added, although the resulting proportion in the second and succeeding reactors keeps decreasing. The same process of hydrogenation with its exothermic effects occurs, and the reactor effluent must be cooled again in a heat exchanger to get it to the right temperature for the next reactor.

The overall conversion of benzene to cyclohexane is nearly 100%, but the effluent from the last reactor still has plenty of hydrogen. To

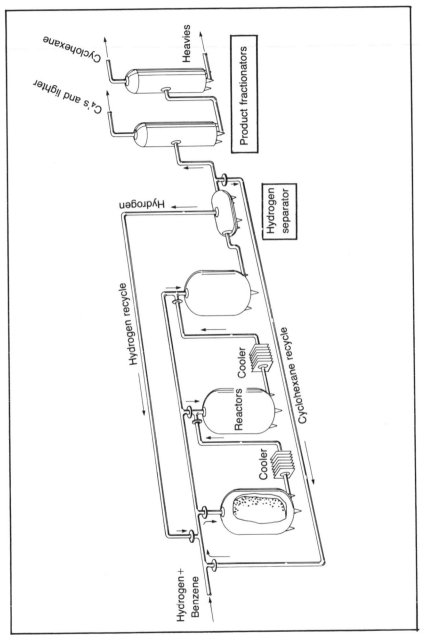

Fig. 4-2 Cyclohexane plant

facilitate the hydrogenation reaction, hydrogen is usually kept in excess. The effluent is passed through a *flash drum*, where the pressure drops. Then the hydrogen flashes out of the product and is recycled to the feed. The remaining effluent is fractionated as a final cyclohexane purification step. Since reaction conditions in the process cannot be controlled perfectly, some of the benzene feed and whatever hydrocarbons come along with it are converted to miscellaneous compounds, including butanes and lighter gases.

A cyclohexane stream also is recycled to the feed and performs an important function. It acts as a *heat sink* or a sponge, diluting the exothermic effect of the hydrogenation reaction and keeping the temperature down. At temperatures higher than 450°F, the decomposition of cylohexane to other compounds (gas for example) increases rapidly.

In summary, the key variables in this process are temperature control, excess hydrogen, sulfur content of the feeds (0%), and catalyst activity. Conversions are typically 99.5%.

Commercial Aspects

Cyclohexane is a colorless, noncorrosive liquid having a very pungent odor. It's flammable, like any naphtha product, and it is shipped in tank cars or trucks, barges, and drums. In commerce, cyclohexane is usually traded on the basis of technical grade (either 95 or 99% purity) or solvent grade (85% minimum purity).

Cyclohexane is used primarily to make caprolactam, adipic acid, and to a lesser extent, hexamethylene diamine (the starting materials for Nylon 6 and 66, synthetic fibers and resins). Nylon-fiber markets include the familiar applications: hosiery, upholstery, carpet, and tire cord. Nylon resins are engineering plastics used mainly to manufacture gears, washers, and similar applications where economy, strength, and a surface with minimal friction are important. Minor cyclohexane uses include industrial solvent applications, such as cutting fats, oils, and rubber. It also is used in paint remover.

V.

OLEFIN PLANTS, ETHYLENE, AND PROPYLENE

Anything that can happen, will happen.

Murphy's Third Law

The big daddy of the petrochemical industry is the olefin plant. The vintage of this process dates before the Forties. Olefin plants are a well-spring of the industry's basic building blocks—ethylene, propylene, butylenes, butadiene, and benzene. The recently built olefin plants are huge. The so-called world-scale plant (the size that achieves whatever is currently considered full economies of scale) is larger than many medium-size refineries. Capacity is no longer measured in millions but billions of pounds per year.

Olefin plants have more than one alias. (One is even fraudulent.) They are variously called ethylene plants (after their primary product); *steam crackers* (because the feed is usually mixed with steam before it is cracked); or _____ crackers, where the blank space is the name of the feed (ethane cracker, gas oil cracker, etc.). Olefin plants are sometimes referred to as ethylene crackers, but that term is the misnomer. Ethylene is not cracked but rather is the product of cracking.

Since ethylene is such a simple molecule, , it stands to reason that many kinds of hydrocarbons can be cracked to form it. The commercial feedstocks in contemporary use range from ethane to heavy gas oil and include propane, butane, naphtha, and light gas oil. The earliest commercial olefin plants of any size were designed to use ethane and propane. As you'll see, these feeds produce a high yield of ethylene and propylene. Technology was well developed by the 1970s, when a major portion of new ethylene capacity was designed to handle the heavy liquids, naphtha and gas oils. The yield of ethylene and propylene from these units was much lower, but the C_4 and aromatic byproducts were valuable.

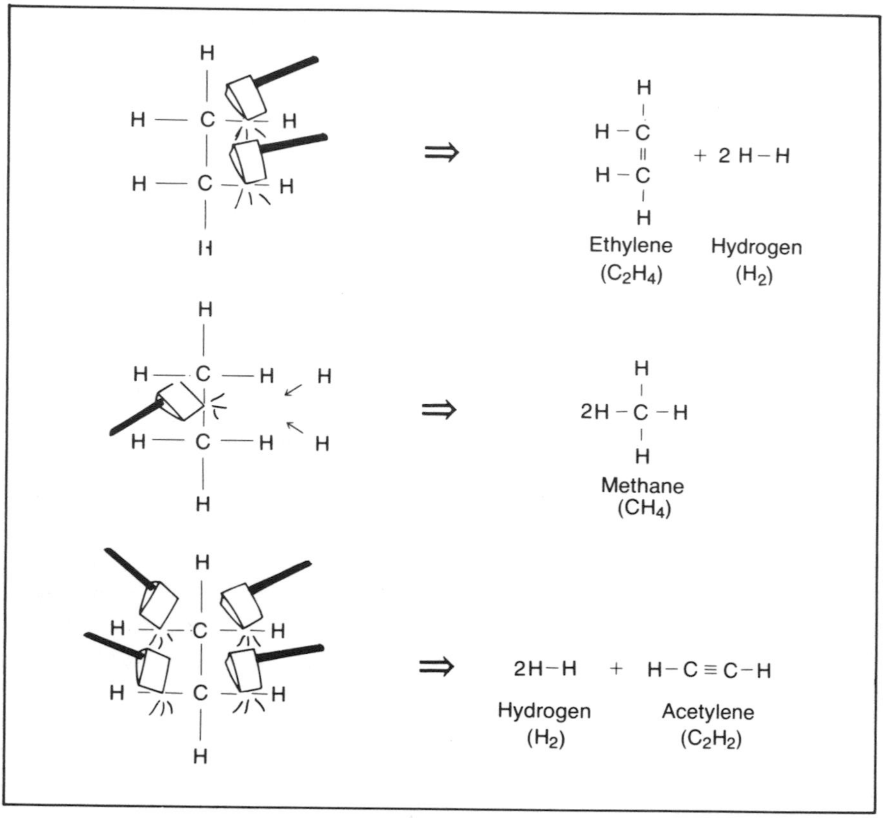

Fig. 5–1   Ethane cracking

Process Description

Ethane and propane cracking is simpler than heavy-liquid (gas oil or naphtha) cracking and should be considered first. When ethane is heated to 1,700°F or more, one of two basic reactions can occur: either carbon-hydrogen or carbon-carbon bonds can split. There is a popular adage in ethylene plant lore, "Anything that can happen, will." The product from ethane cracking depends on which bond is cleavered. Ethylene forms from carbon-hydrogen fractures. Methane comes from carbon-carbon cleavage and hydrogenating the resulting methyl radical. Even acetylene and hydrogen can be formed and survive. However, if the olefin plant is run properly, the predominant yield will be ethylene.

Propane cracking is a little more complicated because there are more combinations and permutations of possible fractures. Not only is there ethylene and methane in the outturn, but also propylene and, surprisingly, ethane. Cracking propane at a carbon-hydrogen bond can give you propylene. Cracking at the carbon-carbon bond produces ethane and methane (after the methyl and ethyl radicals pick up hydrogens) or ethylene if both react at the same time. Ethane also forms as a second-round draft-choice when two methyl radicals find each other instead of hydrogen. Remember, "anything that can happen . . . ."

When naphtha or gas oil is cracked, limitless combinations are possible. Naphthas are made of molecules in the C_5–C_{10} range; gas oils from C_{10} to perhaps C_{30} or C_{40}. The structure includes everything from simple paraffins (aliphatics) to complex polynuclear aromatics. Under the right operating conditions, the naphtha or gas-oil cracking process can be controlled to produce smaller but still acceptable amounts of ethylene and propylene than those from ethane. Table 5–1 shows examples of olefin plant yields.

TABLE 5–1

Olefin Plant Yields

Feed	Pound/Pound of Feed				
	Ethane	Propane	Butane	Naphtha	Gas oil
Yield:					
Ethylene	0.80	0.40	0.36	0.23	0.18
Propylene	0.03	0.18	0.20	0.13	0.14
Butylene	0.02	0.02	0.05	0.15	0.06
Butadiene	0.01	0.01	0.03	0.04	0.04
Fuel gas	0.13	0.38	0.31	0.26	0.18
Gasoline	0.01	0.01	0.05	0.18	0.18
Gas oil	—	—	—	0.01	0.12
Pitch	—	—	—	—	0.10

With naphtha or gas-oil cracking, many other hydrocarbons, including the BTX, show up as byproducts. In the C_4 and C_5 ranges, a new breed called *conjugated dienes* appear. These are molecules with more than one set of double bonds. The commercially important ones are butadiene (CH_2=CH—CH=CH_2) and isoprene $\left(\begin{array}{c} CH_2{=}CH{-}C{=}CH_2 \\ | \\ CH_3 \end{array} \right)$, both of which are important materials for making synthetic rubber.

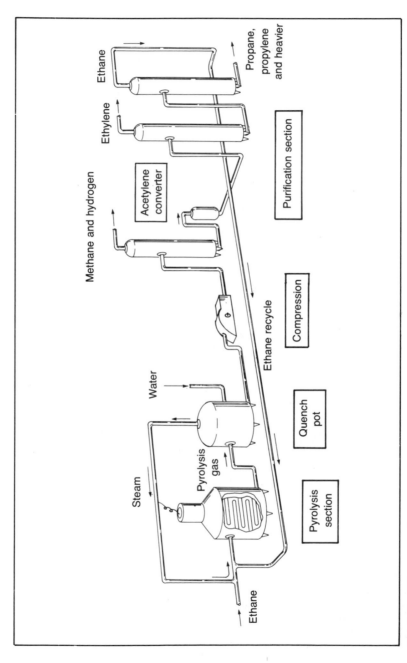

Fig. 5-2 Ethane cracker

The Hardware

Olefin plants have two main sections: pyrolysis or cracking and purification or distillation. Fig. 5–2 shows an ethane cracker that has the simplest purification section. The pyrolysis section (from the Greek, *pyros*, fire) consists of a gas-fired furnace in which the cracking takes place. Ethane is pumped through a maze of tubes where it is heated to about 1,500°F and cracks. (The ethane never contacts the fire directly; instead it stays inside the tubes. Otherwise, it can ignite.)

Ethane is pumped through the pyrolysis section at a very high rate. Residence time of any individual molecules is a few seconds or less. This rapid rate is required to keep the cracking process from running away (operating out of control), resulting in the ethane cracking to methane or coke (carbon) and hydrogen. To control cracking further, the ethane is mixed with steam before it is fed to the furnaces. Steam has two beneficial effects. It lowers the temperature necessary for the cracking process, reducing the fuel bill and the amount of methane and hydrogen that form. Second at the lower temperature, fewer coke deposits form on the inside of the furnace tubes. That saves having to shut down the furnaces for *decoking* so often, a step needed to prevent clogging and cold spots. Coke deposits act as an insulator, and keep the ethane from heating to the right temperature for effective cracking.

As soon as the hot *effluent* (the gases coming from the reaction section) leaves the cracking furnace, it enters a quench pot. The gases from the furnace are so hot they continue to crack, just as a steak continues to cook after you take it off the grill. Therefore, the gases are immediately hit with a stream of cooling water. Heat transfers from the gases to the water, and the water turns into steam. This steam is later separated and is used as the steam mixed with the fresh, incoming feed to the furnaces. At this point, the cracked gases typically consist of a mixture with the following composition:

	Weight %
Methane and hydrogen	8
Ethylene	48
Ethane	40
Propane and heavier	4

You can see that only 60% of the ethane has been cracked; 40% remains in the cracked gas stream. Therefore, part of the purification section is dedicated to separating the ethane, so it can be fed back to the furnaces again. This arrangement is called *recycling to extinction*. Although the pyrolysis section only makes 48% ethylene, recycling resalts in combined pyrolysis/purification yields as follows:

	Weight %
Methane and hydrogen	13
Ethylene	80
Propane and heavier	7

No ethane is produced. These recycled yields can be computed by simple arithmetic. Take the once-through yields; drop the 40% of ethane and divide the other amounts by 0.60 (60%). The products add up to 100%.

In the purification section of an ethane cracker, gas can be handled in one of two ways. In order to fractionate the streams, they must be liquefied. Since they are all light gases, liquefaction can be done by increasing the pressure in a compressor or by reducing the temperature in something called a *cold box*. The ethane cracker in Fig. 5–2 shows the compressor option. In both processes, the streams must be cooled to assure liquefaction.

Downstream of the compressor is a series of fractionators (generally the tallest towers in an ethylene plant) that separate the methane and hydrogen, the ethylene, the ethane, and the propane and heavier streams. These towers have the heavy metallurgy to handle the pressures, and they are insulated to maintain the low temperatures. An acetylene converter is also included. Trace (very small) amounts of acetylene in ethylene can really clobber some of the ethylene derivative processes, particularly when manufacturing polyethylene. Therefore, the acetylene is treated with hydrogen over a catalyst to convert it to ethylene.

It may seem unusual that an ethane cracker has propane and heavier feeds in the outturns. There are two reasons. The ethane used as feed is rarely pure and generally has a small percentage of propane and heavier in it. Also, some of the heavies are actually formed in the frantic scramble of free alkyl radicals during the cracking process.

Heavier Feeds. As the feeds to the olefin plant get heavier, the pyrolysis section changes size and the purification becomes more extensive. As you saw in Table 5–1, the yield of ethylene from the heavier feeds is much lower than from ethane. In the extreme, the yield from gas oil is only about 18%. That means to produce the same amount of ethylene on a daily basis, the gas-oil furnaces must handle nearly five times as much feed as ethane furnaces. Some of the factors that a design engineer worries about are the size of the tubes necessary to heat this much

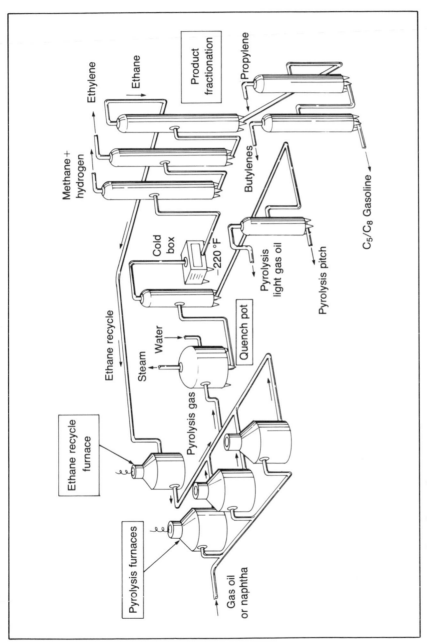

Fig. 5–3 Heavy liquids cracker

feed, the best residence times for each kind of feed, and the best pressure/temperature/steam mixture conditions.

The separation section of a gas-oil cracker looks like a small refinery, as you can see in Fig. 5–3. In addition to the fractionators and treaters in the purification section of the simpler ethane cracker, other facilities separate the heavier coproducts. In the front end of the separator units in Fig. 5–2, the cold-box option for handling liquefication is shown. Temperatures as low as $-220°F$ are achieved in this super-refrigerator. At those low temperatures, freon will not do the job. Liquid air, methane, ethylene, or ammonia are often used as the refrigerant in much the same way freon is used in an air conditioner.

Several new streams are introduced in Fig. 5–3. The C_4 stream is a combination of butanes, butylenes, and butadiene. Depending on the commercial interest, this mixture can be processed further to separate the individual streams. The C_5^+ gasoline stream, usually called *pygas*, is typically given a mild hydrogenation step. Some of the molecules are very reactive olefin and diolefin (two double-bond) structures that are bad actors in gasoline. They form gums and lacquers in car engines. When these elements are removed, pygas becomes a high octane gasoline component because of its aromatics content (*See* chapter II). It also can be processed in an aromatics recovery unit, but even with the benzene removed it makes good gasoline.

Fuel oils from olefin plants also have a high aromatic content. As a result, the burning characteristics of pyrolysis gas oil and pitch are poor. They are smoky, sooty gum formers, and they tend to be more viscous. Also, because of their polynuclear aromatic content (same definition as in Fig. 2–1), these fuels are suspected carcinogens—an unsavory set of products.

In most olefin plants cracking anything heavier than ethane, there is a furnace designed to handle the ethane recycle stream. The plant in Fig. 5–3 shows three heavy-liquid furnaces and one ethane furnace. Since an alternative use for ethane is usually refinery fuel, economics often dictate recovering and cracking this product.

Process Variables

Despite the abundance of analysis in the technical journals on the subject, there really isn't much flexibility in changing the yields in olefins plants. The yield of each of the coproducts moves in a different direction as the pressures, temperatures, and residence times in the furnaces are changed. The fluctuations of the economic values of the byproducts often result in little incentive to effect yield changes.

More significant, however, are the changing values of the feedstocks. In many plants or companies the design allows substituting one feed for another, say, ethane for propane or naphtha for gas oil. In these cases, plant operations respond to the markets for feedstocks and products and reflect themselves in the changing yields implied by Table 5–1.

Ethylene

Ethylene is a colorless gas with hardly any odor. It turns from liquid to gas (boils) at −155°F. It burns readily in the presence of oxygen with a luminous flame. In fact, ethylene made coal gas very useful as a gaslight fuel. The other components in coal gas don't give off nearly as much light when burned alone. Natural gas lamps or propane/butane lanterns must be fitted with mantels to reduce the oxygen available, permitting only partial oxidation to carbon monoxide. However, burning ethylene does not require a mantel.

The logistics of ethylene are tough. Since it's a light gas, high pressures or extremely low temperatures are required to handle it as a liquid. Very little ethylene is transported by truck, and even that must be done under special permit. On long hauls a trucker generally must vent off some of the ethylene to keep the remainder cool enough to be contained at a reasonable pressure.

Venting off ethylene (or any gas) is a cooling process. Ever wet your finger and stick it in the air to see which way the wind was blowing? Wind was coming from whichever side of your finger was cool. The moisture on your finger was vaporizing, and the reaction was a cooling process. For the moisture to go from liquid to vapor, it must pick up heat from its surroundings (your finger). Similarly, when ethylene is vented from a truck, it takes heat from the remaining ethylene. If the temperature is not controlled, the pressure of the ethylene increases dangerously as the ethylene warms up to ambient temperature.

Ethylene trucks are expensive to build and operate. The fuel, the ethylene lost, and the energy needed to liquefy the ethylene are costly factors. Therefore, most ethylene is transported by pipeline, and most consumers are located in proximity to the producers. Although the operating costs of a pipeline are low, the capital costs are high. Thus, ethylene transportation, like most other aspects of the product, is capital intensive.

Ethylene pipelines operate more like natural gas than petroleum pipelines. That is, the ethylene moves as a gas, not as a liquid because of its critical temperature. There's an interesting physical phenomenon involved here. Every gas has a critical temperature. If you keep the gas

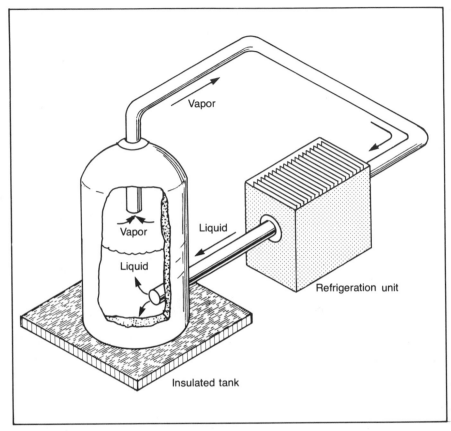

Fig. 5–4   Cryogenic ethylene storage

above that temperature no matter how much you increase the pressure, the gas will not liquefy. This characteristic is associated with atomic structure. The critical temperature for ethylene is 48.6°F. Ethylene pipelines are usually buried about 10–15 feet below ground level, so the surrounding temperature is always 60–70°F. Ethylene, then, can be pumped at very high pressures, 700–800 psi. At those temperatures it is a very dense gas—nearly as dense as a liquid, but still a gas.

Ethylene storage also is an expensive proposition. For small volumes like transfer tanks in a chemical plant, steel storage can be used. But the pressure requirements at normal temperatures demand expensive heavy-duty, thick steel vessels. Storing any amount beyond 100,000 pounds warrants cryogenic units (from the Greek *kryos*, "cold" and *gen*, "to bring forth"). Cryogenic tanks are much lighter and cheaper than steel tanks. They can be used because the ethylene is supercooled

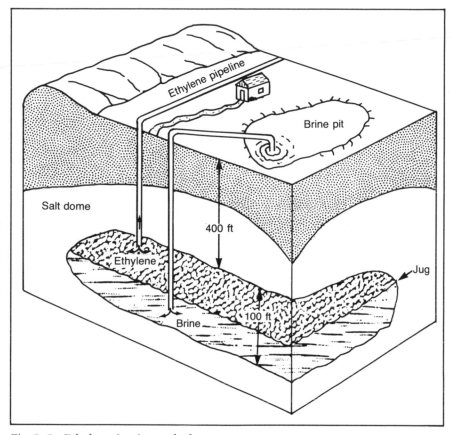

Fig. 5–5   Ethylene jug in a salt dome

below the critical temperature. Under this condition the ethylene is liquid, and very little pressure is needed to keep this liquid from vaporizing. The operating cost of cryogenic tanks is high. Although thick insulation is used around the tank, some heat leaks into the ethylene. To keep the ethylene below the −155°F boiling point, some of the vapor is drawn off the tank and is passed through a refrigeration unit to be liquefied. Then it is returned to the tank. Circulating this stream faster or slower through the refrigeration unit keeps the liquid ethylene temperature in balance with the change in temperature outside the tank.

For large inventories of ethylene (in the millions of pounds), underground storage has been very cost effective. This method usually takes the form of *caverns* mined in rock, shale, or limestone or *jugs* leached out of large underground salt domes (as shown in Fig. 5–5).

In a jug (the more common facility), ethylene is pumped in and out by letting brine (salt water) in and out of the jug. Like any other hydrocarbon/water combination, the ethylene and water do not mix. Therefore, the water acts as a pressuring agent on the ethylene. Jugs or caverns are generally located a few hundred feet below ground level. Ground temperature is constant (65–70°F) and is always above the critical temperature. The weight of the water in the stand pipe is enough to keep the ethylene compressed and at a normal—not cryogenic—temperature for pipeline transport.

Salt-dome construction is cheaper than mined-cavern storage, which in turn is much less per pound of ethylene than cryogenic and pressure storage systems.

The chemical uses for ethylene prior to World War II were limited for the most part to ethylene glycol and ethyl alcohol. After the war, the demand for styrene and polyethylene took off, stimulating ethylene production and olefin plant construction. The list of chemical applications for ethylene reads like the "What's What" of petrochemicals. Polyethylene, styrene, vinyl chloride, ethylene oxide, ethylene glycol, ethyl alcohol, vinyl acetate, alpha olefins, and linear alcohols are some of the more common commercial derivatives of ethylene. The consumer products derived from these chemicals are found everywhere, from soap to construction materials to plastic products.

Propylene

Propylene, like ethylene, is a colorless gas at room temperature. It is as flammable as LPG (liquefied petroleum gas or propane). In fact, propylene can be used as a substitute or supplement to LPG. The fuel characteristics are almost indistinguishable.

Propylene is traded commercially in three grades: refinery, chemical, and polymer. The difference is determined by the ratio of propylene to propane in the stream. Refinery-grade propylene usually runs about 50–70% propylene; chemical grade, 90–92%. Polymer grade is at least 99%. In each case the remaining percentage is almost all propane.

These three grades have been named very practically. Refinery-grade propylene streams are generally byproducts of a refinery cat cracker, and the propane/propylene ratio is determined by the way the cracker is run to make gasoline, not propylene. Chemical-grade propylene is usually produced in a naphtha or gas-oil cracker. The ratio of propylene and propane is 92 : 8 for most operating conditions. Some applications, particularly those to manufacture polymers like polypropylene, require very pure propylene feed. Polymer-grade is made by fractionating one of the less-pure propylene streams, refinery or chemical grade.

The logistics of propylene are more conventional than ethylene, but they are still expensive. Ethylene is like natural gas, and propylene is almost identical to LPG (liquefied petroleum gas or propane). At room temperature, propylene must be kept in a pressurized container to keep it from evaporating. It boils at $-54°F$, so cooling it to keep it liquid is expensive.

Propylene is moved in equally large volumes by pipeline, tank car, and truck. All three modes handle propylene as a liquid, operating at pressures of about 200 psi. The storage facilities for propylene are much the same as those for ethylene: underground jugs or caverns and cryogenic units. In addition, since the operating pressures to keep propylene liquid are lower, storage units in the form of steel spheres (typically 5–10 million pounds) and cylinders or bullets (200–500 thousand pounds) are prevalent.

Unlike ethylene, more propylene is produced than is needed for the chemical industry. As a consequence, there has always been a large amount of propylene in manufacturing gasoline. The most popular process has been alkylation, in which a high-octane, C_7 hydrocarbon is made by reacting propylene with isobutane in the presence of sulfuric or hydroflouric acid. The product is called propylene alkylate and has an octane number of about 96, so it is a good gasoline-blending component.

The larger chemical uses for propylene are to manufacture polypropylene, propylene oxide, isopropyl alcohol, cumene, and acrylonitrile. Many consumer products that you're familiar with show up everywhere, such as plastics in automobile parts, polypropylene rope, rubbing alcohol, epoxy glue, and carpet.

VI.

THE C$_4$ HYDROCARBON FAMILY

If you cannot get rid of the family skeleton,
you may as well make it dance.

George Bernard Shaw (1856–1950)

The first serious notice of C$_4$ hydrocarbons came with the development of refinery cracking processes. When catalytic cracking became popular, refiners were faced with disposing of a few thousand barrels per day of a stream containing butane, butylenes, and small amounts of butadiene. At first, it was all burned as a refinery fuel. The advent of the alkylation plant resulted in most of the butylenes being converted to alkylate, a high-octane gasoline blending component.

During World War II, the Japanese cut off U.S. access to sources of natural rubber, and synthetic rubber manufacturing blossomed. The C$_4$ stream was a direct source of butadiene, a synthetic rubber feedstock. Additional butadiene also was manufactured by dehydrogenating butane and butylene.

In the 1950s, olefin plants cracking ethane, propane, and butane began to produce modest amounts of byproduct C$_4$ hydrocarbon streams. Later in the 1960s and 70s, gas-oil- and naphtha-based olefin plants rivaled refineries in the size of the C$_4$ streams being produced.

A typical C$_4$ hydrocarbon stream from a gas-oil or naphtha cracker has the following composition:

Isobutane	5%
Normal butane	5%
Butadiene	42%
Isobutylene	18%
Butene-1	18%
Butene-2	12%

The terms *butene-1* and *butene-2* are usually used in the petrochemical business instead of the term *normal butylene* because they are more descriptive. To complicate matters, there are two kinds of butene-2: *cis*-butene-2 and *trans*-butene-2. In Table 6–1, you can see that the difference between these three butenes is the location of the double bond.

Table 6–1

Characteristics of the C$_4$ Hydrocarbons

Configuration	Name	Boiling Temperature, °F
$\underset{\displaystyle CH_3-\overset{\displaystyle \overset{CH_3}{\mid}}{CH}-CH_3}{}$	Isobutane	10.9
$CH_3-\overset{\overset{CH_3}{\mid}}{C}=CH_2$	Isobutylene	19.6
$CH_3-CH_2-CH=CH_2$	Butene-1	20.7
$CH_3-CH=C=CH_2$	1, 2 butadiene	51.4
$CH_2=CH-CH=CH_2$	1, 3 butadiene	24.1
$CH_3-CH_2-CH_2-CH_3$	Normal butane	31.1
$\underset{H}{\overset{CH_3}{}}\!\!\diagdown\!\!C\!=\!C\!\diagdown\!\!\underset{CH_3}{\overset{H}{}}$	Trans-butene-2	33.6
$\underset{H}{\overset{CH_3}{}}\!\!\diagdown\!\!C\!=\!C\!\diagdown\!\!\underset{H}{\overset{CH_3}{}}$	Cis-butene-2	38.7

Butene-1 has a double bond at the end (number 1) position; butene-2 at the middle (number 2). The difference between *cis-* and *trans*-butene-2 is more subtle. The methyl groups in *trans* are across from each other, on opposite sides of the fence. In *cis* they are next to each other or on the same side of the fence. Believe it or not the position of the methyl group determines how the molecule behaves. For instance, check the boiling points in Table 6–1. This table also shows the difference between the two butadienes.

Processing

There are a dozen different ways to handle the C$_4$ stream in a petrochemical plant. Simple fractionation won't do it because the boiling points of the components are so close. Generally the first step is to remove the butadienes by extractive distillation. Isobutylene can be removed by a reaction with methanol (CH$_3$OH) to make MTBE (methyl tertiary butyl ether) or by polymerization. Then butene-1 can be removed by selective absorption or by distillation. The butene-2 components remain with iso- and normal butane, which are generally used as feed to an alkylation plant.

A typical C_4 processing scheme is shown in Fig. 6–1. However, most chemical plants do not have all of these facilities. Furthermore, some plants have processes to convert butenes to butadienes; others do just the opposite. The best way to sort out the options is to discuss each byproduct individually.

Butadiene

Butadiene is used as a feedstock for synthetic rubber, elastomers, and fibers. It has grown to be a major petrochemical building block and commodity. Butadiene is a colorless gas at room temperature, but it is normally handled under pressure or refrigerated as a liquid.

Butadiene from olefin plants is the base-load supply simply because it is coproduced with the other olefins. There's not much decision on whether to produce it; it just comes out. Olefin-plant butadiene is about 75% of the total butadiene produced in the U.S. The swing supply butadiene, or that produced on purpose, is made by catalytically dehydrogenating (removing hydrogen from) butane or butylene.

Process Description

The extractive distillation process to remove butadienes from the C_4 stream uses a solvent that reduces the boiling point of the byproducts. The C_4 is fed to the middle of a fractionator, and a high boiling-point solvent is fed in at the top. As it works its way down, the solvent strips the butadiene from the C_4 vapor moving upward. The solvent and butadiene come out the bottom and can be split easily in a second column. Two popular solvents are n-methyl pyrrolidine (NMP) and dimethyl-formamide (DMF).

The dehydrogenation process for making butadiene from butane or butylene involves passing the feed over a catalyst at about 1,200°F and at a reduced pressure. The catalysts used are ferric or chromic oxide, calcium nickel phosphate, or chromia-alumina. After one pass, the effluent usually goes through the extractive distillation process, which separates the butadiene from the recycle streams.

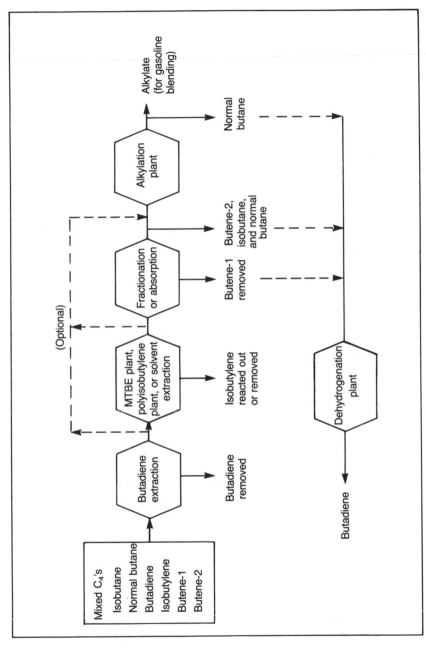

Fig. 6–1 Typical C₄ processing scheme

End-Use Markets

Butadiene's two double bonds make it very reactive. It readily forms polymers, reacting with itself to form polybutadiene. It also is used as a comonomer to make styrene-butadiene rubber (SBR), polychloroprene, and nitrile rubber. Most of the polymers from butadiene end up on highway vehicles: truck and car tires, hoses, gaskets, and seals. Some of the nonrubber applications are adiponitrile and hexamethylenediamine, to make Nylon 66 for carpet fiber and backing. Other nonrubber applications are styrene-butadiene latexes for paper coatings and carpet backing and acrylonitrile-butadiene-styrene (ABS) resins for plastic pipe and automotive/appliance parts.

Isobutylene

The isobutylene in the C_4 stream generally goes to one of four places: an MTBE plant, a polymerization process, a solvent-extraction process, or a refinery alkylation plant. The first three are methods of removing isobutylene from the C_4 stream. The fourth is a default because the isobutylene follows the normal butenes to a process for making gasoline.

The MTBE plant turns out a high-octane gasoline blending component. The process is attractive because the entire C_4 stream can be fed to it, and the isobutylene is selectively reacted out. Similarly, the polymerization process pulls the isobutylene out of the C_4 stream. The polyisobutylenes are used mainly as viscosity index improvers in lubricating oils and as caulking and sealing compounds. Some of the low molecular-weight polyisobutylenes are particularly suited for construction operations because they don't solidify. These compounds remain tacky fluids. When properly formulated with such materials as clay fillers, they take on the properties of a sticky, putty-like substance.

The segregated isobutylene stream from the solvent extraction process has numerous applications besides the polyisobutylenes just mentioned: butyl rubber, alkylated aromatics, oxo alcohols, tertiary butyl alcohol, and di- and tri-isobutylenes.

Process Descriptions

In an MTBE plant, isobutylene is selectively reacted with methanol over a catalyst to produce methyl tertiary butyl ether.

$$CH_3-\underset{\underset{\displaystyle }{}}{\overset{\overset{\displaystyle CH_3}{|}}{C}}=CH_2 + CH_3OH \longrightarrow CH_3-\underset{\underset{\displaystyle CH_3}{|}}{\overset{\overset{\displaystyle CH_3}{|}}{C}}-O-CH_3$$

The conversion occurs in a fixed-bed reactor. The process is exothermic, so cooling coils must remove the heat from the reactor continuously. The effluent contains unreacted feeds and MTBE, and it is fractionated into recycle streams and product.

The isobutylene polymerization process is a low-temperature, catalytic reaction. The type of polymer produced is strongly affected by the reaction temperature. Low temperatures give low molecular-weight polymers, the kind useful in caulking compounds.

A high purity isobutylene stream can be recovered by solvent extraction using sulfuric acid. One problem occurs if any butadiene remains in the stream. Then sulfuric acid will cause it to polymerize. However, if the butadiene has been extracted first, a 99+% isobutylene stream can be recovered.

Butene-1

The demand for high purity butene-1 (called polymer grade) rapidly developed in the 1970s when it became a popular comonomer with ethylene in high-density polyethylene (HDPE) and linear low-density polyethylene (LLDPE). Other minor uses of butene-1 are to manufacture SBA (secondary butyl alcohol), polybutene, and butylene oxide.

The boiling points of butadiene, isobutylene and butene-1 make it impractical to recover a high purity butene-1 stream without first removing the other two. Then butene-1 can be recovered through fractionation. That's still an expensive proposition because the boiling temperatures of isobutane, normal butane, and butene-2 are not that far away. An alternative is molecular-sieve absorption, which works well.

Process Descriptions

The applications of butene-1 require very low levels of isobutylene and butadiene. An extra reactor added to the MTBE plant can get the isobutylene content down from the typical 2% to a 0.2% level. Small amounts of butadiene are removed by hydrotreating the stream over a catalyst, which converts the butadiene to butene-2 and possibly some butane.

The distillation method of separating butene-1 requires a column with over 100 trays and a reflux ratio of about 150:1. (Chemical engineering jargon that means a very tall column with lots of recycle—very energy intensive.)

The second technology is absorption, which uses molecular sieves like those discussed in chapter II on xylenes. Molecular sieves, you will recall, are crystals with millions of pores of a uniform size or shape. In this process, a sieve with pores that will fit only butene-1 is used.

The process runs on a set cycle. First the C_4 stream is fed to a vessel packed with the molecular sieve, and the butene-1 molecules fill up the pores. When the pores are nearly saturated, the feed is cut off. Another liquid, called a *desorbent*, is flushed back through the vessel, and the butene-1 is washed out of the sieves. The desorbent is selected so that after it picks up the butene-1 from the sieve, it can be easily separated from the butene-1 by fractionation. The key, of course, is to use a desorbent with a boiling temperature a good distance from that of the butene-1.

Butene-2

At the end of the line, when all the goodies (byproducts) have been removed, is the alkylation plant. That's where most of the butene-2 goes because there are few chemical applications for this molecule. The only thing good you can say about butene-2 is that it makes a better (higher octane) alkylate than butene-1.

Alkylation is the addition of an olefin to a branch chain hydrocarbon. In a refinery alkylation plant, isobutane is the branch chain; the olefins are propylene and butylenes. The result is a C_7 or C_8 hydrocarbon with good motor gasoline characteristics that is called *alkylate*.

Process Description (Alkylation)

Mixed C_4 or C_3 streams are fed to a reactor, together with any excess isobutane (about a 12 to 1 ratio). The reactors contain cold sulfuric or hydroflouric acid, which acts as a catalyst. Active mixing during a long residence time (15–20 minutes) results in a reaction of the olefins and isobutane. The propane and normal butane are unaffected and float on through the process. Distillation at the end of the plant easily separates the alkylate from the propane, normal butane, and isobutane. The isobutane is recycled to the reactor. Quality-wise, the alkylate made from butenes is better than that from propylene because it has a higher octane rating.

*This is not the end. It is not even the beginning of the end.
But it is, perhaps, the end of the beginning.*

Winston Churchill (1874–1965),
commenting on November 10, 1942
about the victory at Alamein

After five chapters of product and process descriptions, you may be dismayed to learn that it all can be summarized in one diagram. The essential parts are shown in Fig. 6–2 to provide a quick reference.

The petrochemical products from olefin plants are ethylene, propylene, C_4's (butadiene and the butylenes), and a stream containing benzene. Refinery cat crackers produce propylene and C_4's. They also manufacture some ethylene, but often it is not recovered. The propylene from olefin plants is usually chemical grade; from cat crackers, refinery grade. Both can be upgraded to the polymer level.

Refinery cat reformers produce a reformate stream laden with aromatics. That stream, with or without the benzene stream from the olefin plant, can be split in the various processing schemes in the BTX recovery facility. The C_4's also can be split in numerous ways to make petrochemical feeds. Often everything but the butadiene ends up in an alkylation plant, and the benzene can be hydrogenated to cyclohexane directly.

The next five chapters cover the "Mutt and Jeff" petrochemicals. They're grouped together because they come in pairs. You don't have one without the other. This metaphor is accurate for cumene and phenol, ethylbenzene and styrene, and ethylene dichloride and vinyl chloride. It becomes a little strained with ethylene oxide and ethylene glycol and with propylene oxide and propylene glycol because there are other applications for the oxides. However, there's virtually no other use for cumene, ethylbenzene, and ethylene dichloride than to use each to make its mate. Anyway, these ten chemicals are second or third generation derivatives of the basic building blocks and are important commodity petrochemicals.

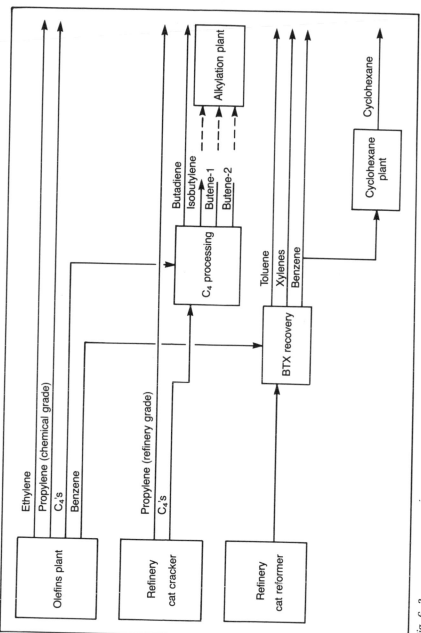

Fig. 6-2

VII.

CUMENE AND PHENOL

Every man serves a useful purpose: a miser, for example, makes a wonderful ancestor.

Laurence J. Peter

The only reason petrochemical companies make cumene now is to use it to make phenol. There are other ways to make phenol, but there are few other commercial uses for cumene.

Cumene

During World War II, isopropyl benzene (more commonly and commercially known as cumene) was manufactured in large volumes for aviation gasoline. By combining a benzene ring and an isoparaffin structure, a very high-octane product was made at a relatively cheap cost. After the war, the primary interest in cumene was for manufacturing *cumene hydroperoxide*. This compound was used in small amounts as a catalyst in an early process of polymerizing butadiene with styrene to make synthetic rubber. It was discovered by accident that mildly treating cumene hydroperoxide with acid resulted in phenol and acetone.

Process Description

The reaction of benzene with propylene produces cumene, but a catalyst must be present to cause the reaction. The chemistry is such that the benzene-propylene bond is at the middle carbon of the propylene molecule, hence, the name *iso*propyl benzene. Note that a hydrogen also transfers from the benzene to the propylene.

The reaction can be carried out with benzene and propylene in either the liquid or vapor phase. The more common process is vapor at about 425°F and 400 psi. The process diagram in Fig. 7–2 shows propylene and benzene fed directly to the reactor. Usually chemical-grade propylene is used because the 6–10% propane diluent does not affect

68

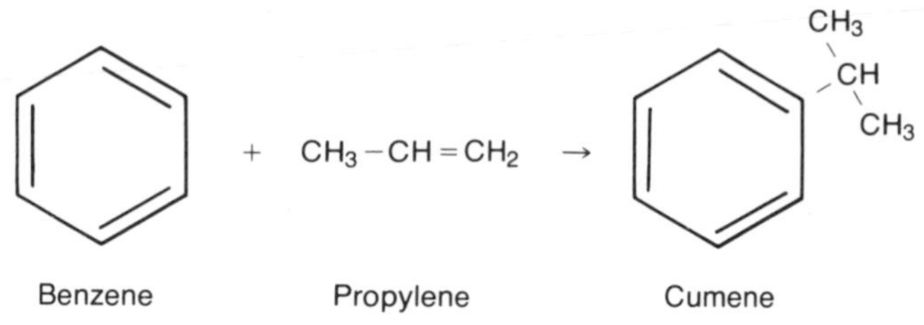

Fig. 7–1　Benzene-propylene route to cumene

the reaction. However, a *depropanizer* must be used for a refinery-grade stream to bring the propane content down to that range.

The reactor is a vessel with supported beds of solid catalyst. Most commercial processes use phosphoric acid deposited on a silica/alumina pellet called *kieselguhr*. Because of the weight of the pellets, supported beds at multiple levels in the vessel are used so the bottom layers will not be crushed.

The reaction of the benzene with the propylene takes place in two vessels for two reasons. First, the reaction is exothermic. In a fixed catalyst bed, the only way to control the temperature is to remove the streams being processed out and cool them down. Second, the other reactor also is a fractionator, venting the unreacted propylene and any propane diluent from the benzene/cumene mix. Excess benzene is always used in the reactors. Since it acts like a heat sponge, the benzene mitigates the rate at which the temperature increases due to the exothermic reaction. Also, this excess benzene helps eliminate some of the undesirable side reactions that can occur, mainly the formation of di-isopropyl benzene (benzene hooking up with two propylenes) and polypropylene.

Therefore, the streams coming out of the reactors are a mixture of the excess benzene and the product cumene. A fractionating column separating the two permits recycling the benzene. A final fractionator separates any miscellaneous compounds accidently formed in the process. Cumene made in this manner is about 99.9% pure. The yield of benzene to cumene is about 95%. The yield based on propylene is a little lower (about 90%), particularly if there is no depropanizer up front to recycle the propane/propylene reactor effluent.

Occasionally you might come across a compound called pseudo-cumene, which is a benzene ring connected to three methyl radicals. This compound is an isomer of cumene known as 1,2,4-trimethyl benzene. Pseudocumene is a starting material for manufacturing *trimelletic*

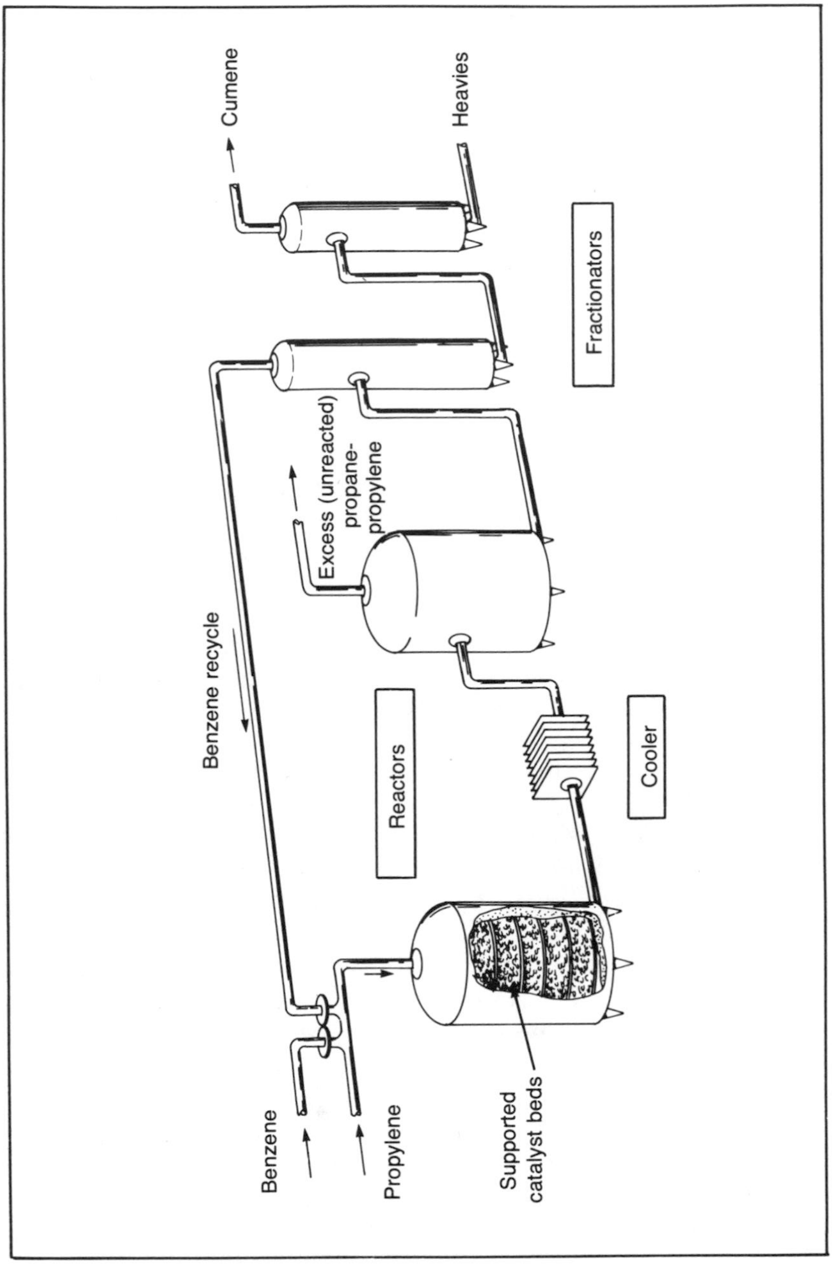

Fig. 7-2 Cumene plant

anhydride, an important ingredient in alkyl resin paints and high-temperature aerospace polyimide resins.

Commercial Aspects

Cumene is a colorless liquid that can be shipped in tank cars, trucks, or drums. The flash point is high enough that it is not considered a hazardous material. The grades used in commerce are technical (99.5% concentration), research (99.9%) and pure (only trace impurities).

Phenol

In the medical field phenol has been used for decades as an antiseptic called carbolic acid and as the preservative of human organs that is named creosote (from the Greek, *kreos,* "flesh" and *sogein,* "to preserve"). Creosote eventually became associated with the wood preservative; however, phenol remains a principal ingredient in this product.

The early sources of phenol were the destructive distillation of coal and methyl alcohol manufactured from wood. In both cases, phenol was a byproduct. Recovered volumes were limited by whatever was created accidentally in the process. Initial commercial routes directly toward phenol production involved reacting benzene with sulfuric acid (1920), chlorine (1928), or hydrochloric acid (1939). All of these were followed by a subsequent hydrolysis step (a reaction with water to get the —OH group) to produce phenol. These reactions required high temperatures and pressures, and they involved multistep processes. Therefore, special metallurgy was needed to handle the corrosive mixtures involved.

In 1952, a technological breakthrough was found: the cumene oxidation route. This process was much cheaper. It quickly proliferated and is now the primary phenol process. Despite the fact that the additional step of making cumene was involved, the less severe operating conditions throughout (pressures, temperatures, acid strength) were sufficient offsets to make the process economically attractive.

Cumene Oxidation Process

This two-step process involves oxidizing cumene to cumene hydroperoxide. Then that compound is broken into phenol and acetone, as

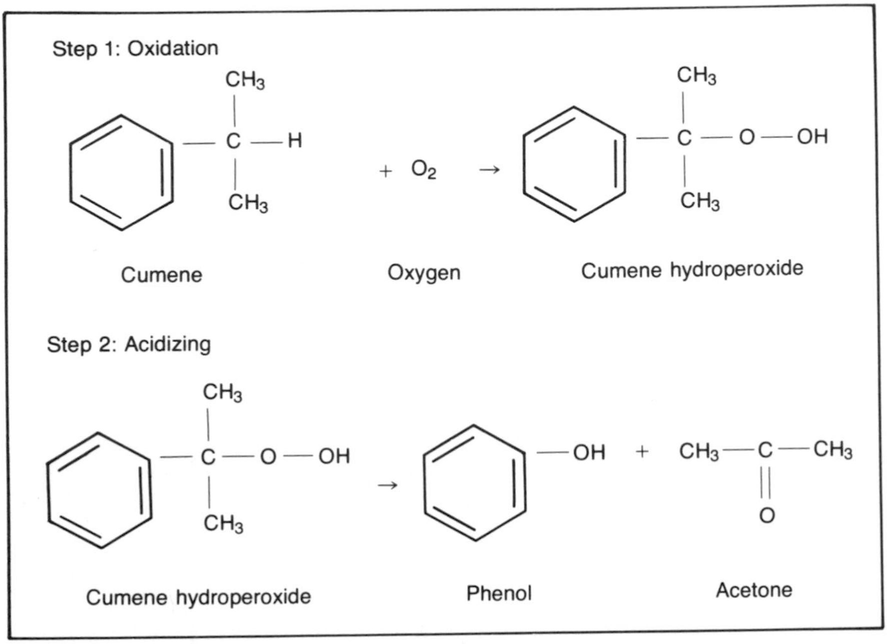

Step 1: Oxidation

Cumene   Oxygen   Cumene hydroperoxide

Step 2: Acidizing

Cumene hydroperoxide   Phenol   Acetone

Fig. 7–3   Cumene-to-phenol process

shown in Fig. 7–3. In the first step, cumene is fed to an oxidation vessel shown in Fig. 7–4. Then it is mixed with a diluted aqueous, sodium-carbonate solution (soda ash with a lot of water). A small amount of sodium stearate is added, and the whole mixture becomes an *emulsion.**

The purpose of the cumene emulsion is to permit good contact of the cumene with oxygen. The oxygen is introduced as air in the bottom of the vessel, and it bubbles through the emulsion. As this happens the cumene converts to cumene hydroperoxide, as shown in step 1 of Fig. 7–3. The chemical reaction (like most oxidations) is exothermic, so it generates heat and is susceptible to a runaway reaction. Rapidly increasing temperatures can increase the reaction rate, which can increase temperatures more rapidly. The presence of the excess water sponges up some of the heat and reduces the risk of runaway.

---

*Emulsions abound in everyday use. Mixing flour and water with meat juices makes the emulsion called gravy. Putting soap powder and water in a washing machine causes the dirt in clothes to be removed and to become suspended in the water—an emulsion. The dirt particles remain in this state, until they are rinsed away by draining the washing machine.

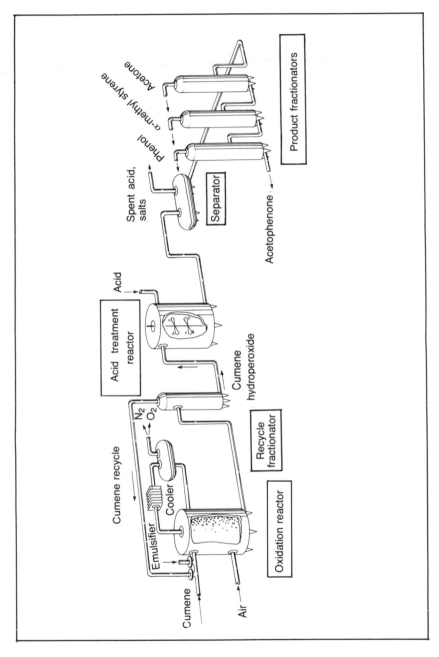

Fig. 7-4 Phenol plant

To control the runaway risk further, the reaction temperature is kept at about 250°F by regulating the reactor flow-through rate. At that temperature only about 25% of the cumene is converted. Therefore, the stream coming out the bottom of the oxidizing vessel is 25% cumene hydroperoxide and 75% unconverted cumene.

At the top of the vessel is the plumbing needed to take the nitrogen from the air, since it passes through untouched. Any excess oxygen also is removed. The bottom stream is fed to a fractionator to split most of the cumene out and to recycle it to the oxidizing vessel. The cumene hydroperoxide, now concentrated to about 80%, is fed to another vessel for the second reaction step. The chemical trick in this one is to chop out everything in between the benzene ring and the — OH group, as shown in step 2 of Fig. 7–3. Diluted sulfuric acid initiates an unusual reaction, involving actual migration of the benzene ring around the cumene hydroperoxide molecule. The process was completely a surprise when it was discovered. Like so many other petrochemical processes, serendipity plays a large role in generating progress.

To facilitate the reaction, the mixture is stirred vigorously. However, this reaction occurs at about 125°F and 25–50 psi, so the conditions here as in the oxidizer are not too severe or expensive.

The effluent from the acid-treatment reactor is about 60% phenol, 35% acetone, plus some miscellaneous cats and dogs (most of which are alpha-methyl styrene and acetophenone). The effluent is passed through a separator where the acid, water, and salts drop out. The balance of the processing is a series of distilling columns that split out the various streams.

Usually the alpha-methyl styrene is catalytically converted back to cumene and is recycled. The acetophenone has some commercial use in pharmaceuticals and at one time was used to make ethylbenzene. A high purity phenol is sometimes made by a crystallization step, since the temperature at which phenol freezes is about 109°F.

Other Routes

A few other commercial phenol processes also should be mentioned here. One involves oxidizing toluene with a cobalt catalyst to give benzoic acid. This process is followed by a carboxylation step to produce salicylic acid, phenol, and carbon dioxide. However, the process has several inherent problems, and it has not proliferated. Another route is to oxidize benzene directly, a chemist's dream. But this method is difficult to control because of the side reactions, and it has not been successful. The third route is a little more popular. Cyclohexane is oxidized to cyclohexonol, which is then dehydrogenated to phenol. Finally, there is the benzene/chlorine process, which hydrolyzes monochlorobenzene to phenol.

The cumene process remains the economic method for the reasons mentioned. Its only drawback is the coproduction of acetone, which is a problem when acetone demand is low and phenol demand is high. What do you do with the excess acetone production? When both phenol and acetone demands are high, the problem often takes care of itself. Then plants that make acetone as a single product can be run at optimum conditions (see chapter XIV).

Commercial Aspects

Uses. The largest outlet for phenol is manufacturing *phenolic resins.* The reaction of phenol with formaldehyde (embalming fluid) gives liquid resins (used extensively as the adhesive in plywood) and solid resins (used as engineering plastics in electrical applications). In powder form the resins can be molded easily and are completely nonconductive. Phenolic plastics can be found in panel boards, switch gears, and telephone assemblies. The agitator in your washing machine is probably a phenolic resin.

Phenol also is used to manufacture several important monomers. *Bisphenol A* is a derivative used to make very strong *polycarbonate* plastics and epoxy resins (the kind you buy in two tubes and mix to make glue). Other applications of epoxy resins include paints, fiberglass binder, and construction adhesives. About half of the *caprolactam* is made from phenol. (The other half comes from cyclohexane.) Caprolactam is the intermediary to Nylon 6. Other miscellaneous derivatives of phenol include nonionic detergents, aspirin, disinfectants (pentachlor phenol), adipic acid (Nylon 66 intermediate), and plasticizers.

Properties. Phenol is a solid at room temperature and is usually handled as a powder. In its pure form it's white, but exposure to sunlight or air will cause it to turn reddish. Phenol is and acts like acid. It burns; it's corrosive; and it has an odor and taste that will knock you over—literally. It is a Class B poison.

Phenol can be shipped in liquid form in lined tank cars or trucks or in galvanized drums. Phenol must be handled in closed systems because it absorbs water from the atmosphere. In powder form, phenol can absorb enough water from the atmosphere to turn itself into a liquid.

The powder form of phenol is usually traded either as a U.S.P. (98% minimum) or as a C.P. or synthetic grade (95% minimum).* In the liquid form, the commercial grades are 90–92% and 82–84% purity.

If the cumene-to-phenol route hadn't been discovered accidentally, it probably would have been developed later on purpose. Because of that early fortune, the process is well-developed and understood.

---

*U.S.P. (United States pure) and C.P. (chemical pure) are nomenclature from the pharmaceuticals industry, the former indicating a grade suitable for human consumption or for manufacturing a consumable.

VIII.

ETHYLBENZENE AND STYRENE

Diogenes the wise crept into his vat
And spoke: "Yes, yes, this comes from that."

Diogenes, Wilhelm Busch (1832–1908)
(inventor of the cartoon strip)

Ethylbenzene is to styrene what cumene is to phenol. Ethylbenzene is produced only so you can make styrene. Most ethylbenzene (EB) is made by alkylating benzene with ethylene, as shown in Fig. 8–1.

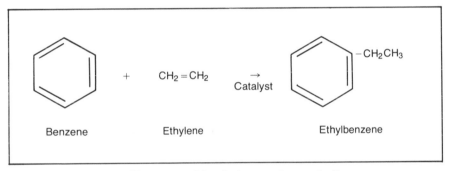

Benzene   Ethylene   Ethylbenzene

Fig. 8–1 Alkylation of benzene with ethylene to form ethylbenzene

A small amount of EB is present in crude and also is formed in cat reforming. You may recall that there is only a 4°F difference between the boiling points of ethylbenzene and para-xylene. Consequently, a superdistillation column is needed for the separation. In process engineering terms, the structure has about 300 theoretical trays, is about 200 feet tall, and has a high reflux ratio. All this is necessary because the EB stream must be quite pure to manufacture styrene.

The Technology

Benzene alkylation is old technology. In 1877, the French chemist, Charles Friedel, with his American partner, James Crafts, stumbled across the technique for alkylating benzene with amyl chloride $(C_5H_{11}Cl)$.

A metallic catalyst, in this case aluminum, was the key. The Friedel-Crafts reaction is classical and today is the principal route for alkylating benzene with the compound ethylene to make EB.

The Friedel-Crafts reaction has one major drawback. It does not stop at the monosubstitution stage. That is, the catalyst works so well, the benzene picks up two, three, or more ethylene molecules, forming di-ethylbenzene, tri-ethylbenzene, or higher polyethylbenzenes (PEB) (Fig. 8–2). Chemically it's easier to alkylate EB than benzene. One way to minimize the problem is to carry out the reaction in the presence of an excess of benzene. When an ethylene molecule is in the neighborhood of one EB molecule and twenty benzene molecules, chances are that the ethylene will hook up with benzene, even though it prefers EB.

The other control variable is to set the operating conditions. Certain temperature and pressure levels and concentrations favor benzene alkylation rather than EB alkylation—not exclusively, but predominantly. In fact these variables can be set to cause some of the di- and tri-ethylbenzenes to give up an ethyl radical to benzene to make EB. This process is called *transalkylation* and is shown in Fig. 8–2.

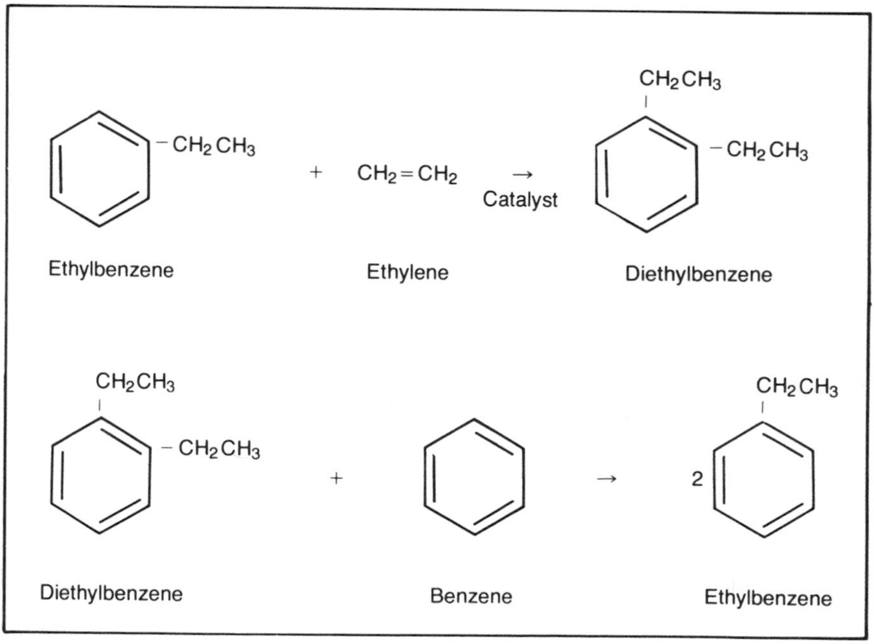

Fig. 8–2 Formation of diethylbenzene and transalkylation back to ethylbenzene

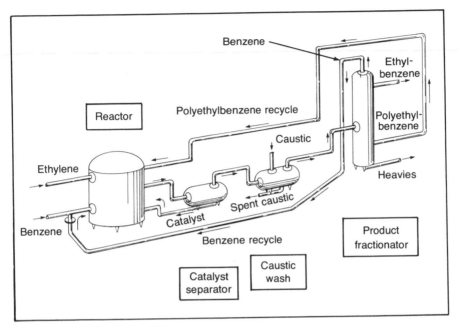

Fig. 8–3   Ethylbenzene plant

Process Description

The hardware to accommodate this reaction can be built for either a liquid or a vapor phase. The liquid operation is more common and is shown in Fig. 8–3. The catalyst is anhydrous aluminum chloride (anhydrous means completely water free). At temperatures of 300–400°F and pressures of 60–100 psi, the reaction time is about 30 minutes. Therefore, the reactor must be large enough to accommodate a long residence time.

Sometimes a catalyst promoter or accelerator, ethyl chloride, is added to the feed to speed up the reaction. The ethyl chloride actually works on the aluminum chloride catalyst, not the reactants. It's like giving a supervisor a bonus. He doesn't do any more work, but he gets more work done.

The effluent stream leaving the reactor is cooled. Then it is treated with caustic (sodium hydroxide) and water to remove the catalyst. The cleaned stream contains about 35% unreacted benzene, 50% ethylbenzene, 15% di-, tri-, and polyethylbenzene, and a small amount of other heavy materials. The EB yields are about 99%, based on the ethylene and benzene feed.*

---

*For a discussion of the difference between yield and conversion, *see* the appendix.

Three fractionators split the four components. Any water remaining after the water wash flows overhead with the benzene. However, it must be removed before the benzene is recycled to the reactor. Usually an azeotropic distillation column is placed before the reactor to dewater the benzene. The presence of any water in the reactor causes many undesirable side reactions, including killing the catalyst.

The polyethylbenzene also is recycled to the reactor. At this point trans-alkylation to ethylbenzene converts this material, so no net PEB is produced. The ethylbenzene from the second column is about 99% pure.

The alternate vapor-phase process is carried out at higher temperatures (485–575°F) and pressures (over 600 psi). This technique uses a solid phosphoric acid or boron trifluoride/silica alumina catalyst in a fixed bed. However, the balance of the process is similar to the liquid-phase method.

Handling

Since most ethylbenzene is used to manufacture styrene, EB plants tend to be captive. Very little EB is traded commercially or transported between facilities. However, a small amount is used as a commercial solvent, mainly as a substitute for xylenes. Ethylbenzene is not toxic like the xylenes, and in certain applications, substituting EB is more practical than using closed systems. Nonetheless, ethylbenzene has the same colorless appearance of the other BTX, as well as that characteristic sickly odor.

Styrene

Styrene should be called one of the basic building blocks of the petrochemicals industry. But you get all mixed up with semantics because it's made from two other basic building blocks, ethylene and benzene. Nevertheless, the rapid growth of styrene after World War II was due to the widespread use of styrene derivatives, principally synthetic rubber and plastics.

Practically all of the styrene today comes from dehydrogenating EB—removing two hydrogen atoms from the ethyl group attached to the benzene ring (Fig. 8–4). Dehydrogenation is done by cracking (like in an olefin plant) in a mixture with steam. In this case, the cracking products are much more limited, primarily because the catalyst is iron oxide.

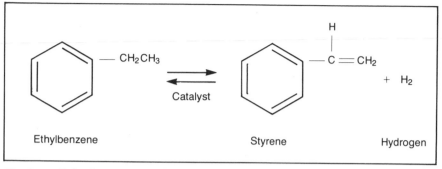

Fig. 8–4 Dehydrogenation of ethylbenzene to styrene

Like all cracking processes, the dehydrogenation step is endothermic —it absorbs heat. Superheated steam mixed with the ethylbenzene provides the heat and performs two other important functions.

1. This steam reduces the pressure at which the reaction occurs. A mixture of ethylbenzene and steam at 1,300°F can be contained at a lower pressure than EB by itself at that temperature. The chemical reaction of EB cracking to styrene and hydrogen is easily reversible; styrene also can hydrogenate to ethylbenzene. Although both always occur to some extent, the primary one is determined by the operating conditions. In this case, higher pressures favor EB formation because ethylbenzene takes up less volume than the corresponding amount of styrene and hydrogen. Conversely, lower pressures form styrene. Therefore, steam mixed with ethylbenzene permits cracking off the hydrogen at lower pressures and encourages the styrene to stay cracked.*

2. The steam also reacts with coke deposits on the iron oxide catalyst, forming carbon dioxide. This reaction gives the catalyst a longer, more active lifetime. The on-stream factor of the styrene plant is extended by reducing the shutdown frequency for catalyst regeneration or replacement.

In this process the conversion rate is about 60% and the yield is about 90%. The 10% yield loss results from cracking one of the carbon-carbon bonds in the EB ethyl group. Consequently, benzene and toluene are the primary byproducts. Another one is polystyrene. The styrene, of course, is very reactive. That's why it is a building block. Although the operating conditions minimize this reaction, some of the styrene molecules do join up to form styrene polymers.

---

*You may have noticed that the chemical equation in Fig. 8–4 has arrows going both directions. That's the chemist's notation for this reversibility.

In this process few benzene rings are detected breaking up. As in the olefin plant, the thermal stability of the benzene ring is demonstrated by its survival of these severe operating conditions, especially the high temperatures.

Process Facilities

The ethylbenzene entering the styrene plant is generally heated to the threshold cracking temperature (about 1,100°F) in a heat exchanger. The counterflow in the exchanger is the effluent from the second-stage reactor, as shown in Fig. 8–5. Because high temperatures are necessary in a styrene plant, conservation plays a big role in plant design. Although this is the only heat exchanger shown, waste-heat recovery is actually an intimate part of the process flow throughout the plant. After heating, the ethylbenzene is mixed with superheated steam and is fed to the first-stage reactor. Both the first- and second-stage reactors are packed with a metal-oxide catalyst on activated charcoal or alumina. Iron oxide, sometimes combined with chromium oxide, or potassium carbonate is commonly used.

The actual reaction occurs at about 1,150°F, but there is a temperature drop in the reactor during dehydrogenation. Reheating in a furnace or exchanger is necessary before the stream is fed to the second-stage reactor for a repeat performance.

The hot effluent is cooled in a heat exchange with fresh EB feed, and it can be cooled even further with water to make steam. Then the stream is sent to separators, where water and the light cracked gases that unavoidably form (H_2, CO, CO_2, CH_4, etc.) are removed. The final product separations are done in a series of fractionators. The ethylbenzene is recycled to the feed line; the polymers, very small in volume, are generally disposed of in residual fuel.

Even at ambient temperatures, styrene is likely to react with itself —slowly, but steadily. Therefore, a small amount of *polymerization inhibitor*, about 10 parts per million (ppm) of para-tertiary butyl catechol, is added to the styrene kept in storage. Since polymerization is promoted by higher temperatures, styrene is usually stored in insulated tanks.

Alternative Routes to Styrene

Very few plants are designed to produce styrene as a coproduct with propylene oxide from ethylbenzene. The steps of this process are

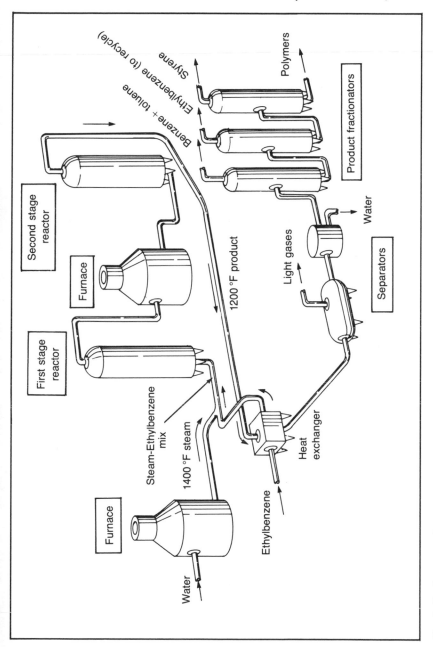

Fig. 8–5 Styrene plant

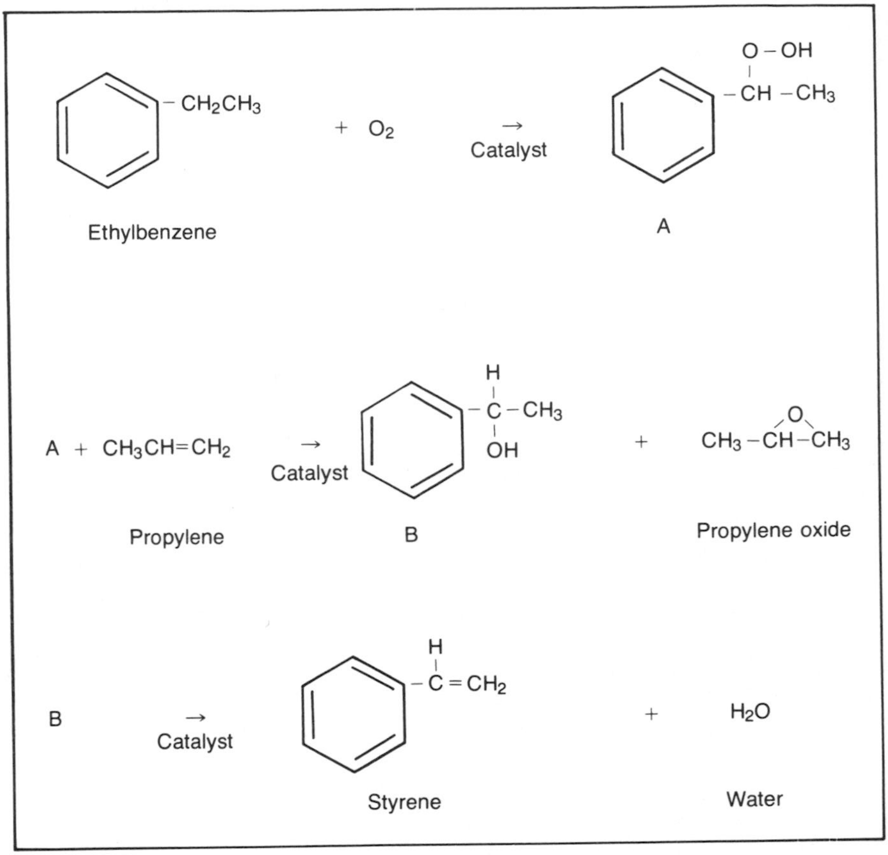

Fig. 8–6 Styrene and propylene oxide by ethylbenzene oxidation

shown in Fig. 8–6. Ethylbenzene is oxidized to hydroperoxide. Also, a catalytic reaction with propylene yields propylene oxide and phenyl methyl carbinol, which dehydrates to styrene. Even fewer plants are designed to oxidize ethylbenzene to acetophenone, followed by hydrogenation to phenyl methyl carbinol and dehydration to, again, styrene.

Commercial Aspects

Uses.   Plastics and synthetic rubber are the major uses for styrene. The numerous plastics include polystyrene, styrenated polyesters, acrylonitrile-butadiene-styrene (ABS), styrene-acrylonitrile (SAN), and styrene-butadiene (SB). Styrene-butadiene rubber (SBR) was a landmark chemical achievement when it was commercialized. Styrene derivatives are found everywhere—in food-grade film, toys, construction pipe, foam, boats, latex paints, tires, luggage, and furniture.

Handling. Styrene is a colorless liquid but tends toward a yellowish cast as it ages. It feels oily to the touch and smells like aromatic compounds. Left alone at room temperature, styrene eventually polymerizes with itself to a clear, glassy solid.

Technical-grade styrene is 99% minimum purity. It is shipped with a polymerization inhibitor in standard tank cars or trucks. However, it has none of the severe handling precautions that benzene has.

IX.

ETHYLENE DICHLORIDE AND VINYL CHLORIDE

Inventing is a combination of brains and material. The more brains you use, the less material you need.

Charles F. Kettering (1876–1958)
President, General Motors

This is the third chapter in a series of three where the products are like pancakes and batter. You can't make pancakes (or vinyl chloride) without making batter (or ethylene dichloride). There's not much else you can do with pancake batter (or ethylene dichloride), and you only use pancakes or vinyl chloride to make something else—breakfast or plastics. Finally, if you will permit this overbearing analogy to be extended once more, making vinyl chloride (or pancakes) from scratch is a lot easier now than it was 40 years ago.

The original manufacturing route to vinyl chloride (VC) did not involve ethylene dichloride (EDC). Instead it was the reaction of acetylene with hydrochloric acid. This process was commercialized in the 1940s but, like most acetylene-based chemistry, gave way to ethylene in the 50s and 60s. The highly reactive acetylene molecule was more sensitive and hazardous and eventually more costly than the rapidly growing ethylene. Most contemporary vinyl-chloride plants use ethylene and chlorine as raw materials.

Vinyl chloride is often called vinyl chloride *monomer* (VCM). The tag-on, monomer (from the Greek *mono*, meaning "one" and *meros*, meaning "part") is a convention to contrast a chemical from its counterpart, the *polymer*. Vinyl is the prefix for any compound that has the vinyl or *ethenyl* radical, CH_2=CH—, in it.

Process Description

Vinyl chloride is made by cracking ethylene dichloride in a pyrolysis furnace much like that in an ethylene plant. That's one of the three reactions, shown in Fig. 9–1, that are involved in the process. The other two

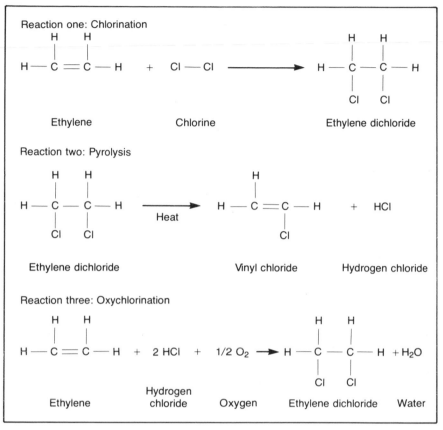

Fig. 9–1   Ethylene dichloride and vinyl chloride reactions

are chlorination and oxychlorination. In these reactions the hydrogen chloride used to make the EDC in reaction three comes from cracking EDC in reaction two. This sounds like a closed circle until you back away and examine it.

The plant with its three reactors is shown in Fig. 9–2. One of the reactors is the pyrolysis furnace in the middle of the figure. At the top of the figure, the basic feeds to the plant are shown: ethylene, chlorine, and oxygen. Ethylene and chlorine alone are sufficient to make ethylene dichloride via the route on the left. This reaction occurs in the vapor phase of a fixed catalyst bed of ferric (iron) chloride at 100–125°F. A cleanup column fractionates the small amount of byproducts formed, leaving an ethylene-dichloride stream of 96–98% purity.

For reaction two, the purified ethylene dichloride is passed through a dryer to remove water. Then it is fed to a pyrolysis unit. The difference between EDC pyrolysis furnaces and those used for ethylene is the catalyst. The tubes in the EDC furnace are packed with charcoal pellets impregnated with ferric (iron) chloride. The ethylene dichloride is pumped through at 900–950°F and 50 psi. The conversion of ethylene dichloride is about 50% and the yield of vinyl chloride is about 95–96%, so not much else is formed. That's a contrast to ethylene manufacturing, especially when cracking the heavy liquids, because those byproducts are abundant.

The hot effluent gas from the furnaces is quenched immediately for the same reasons as ethylene furnace effluent is quenched—to stop cracking at the optimum spot. However, the quench liquid is cool ethylene dichloride, not water.

When EDC cracks, one hydrogen and one chlorine on adjacent carbon atoms are sprung and find each other, forming hydrogen chloride gas. (Why one of each cracks off, and not two hydrogens or two chlorines is another mystery of atomic physics.) The cooled effluent is fractionated into three streams: hydrogen chloride, ethylene dichloride, and vinyl chloride (which is sent to storage). The ethylene dichloride, which is the uncoverted pyrolysis feed plus the substance added in the quench pot, is recycled to the EDC cleanup column. The hydrogen chloride would be a disposal problem, but it is pumped to the oxychlorination stage for EDC production (reactor three in Fig. 9–1).

Oxychlorination takes place in the vessel shown in the upper right of Fig. 9–2. The reactor is packed with a cupric (copper) chloride catalyst. Three feeds (the hydrogen chloride, oxygen in the form of air, and ethylene) are reacted at 600–800°F, to form ethylene dichloride and water. The reaction effluent is then piped to the other EDC stream heading for the cleanup fractionator. There, it commingles with the EDC stream from reaction one and the recycle stream from the VC cleanup fractionator.

Therefore, the two recycle streams (hydrogen chloride and EDC) are treated very differently. The ethylene dichloride is just cleaned up, but the hydrogen chloride is chemically reacted to EDC. Considerable attention must be paid to balancing the flows around this plant. There also are surge tanks in the plant that are not shown in Fig. 9–2. However, they can quickly fill up, potentially causing the need to shut down one of the reactions. Starting and stopping any of the reactions tend to be a problem because end products may be below specifications and because energy is wasted.

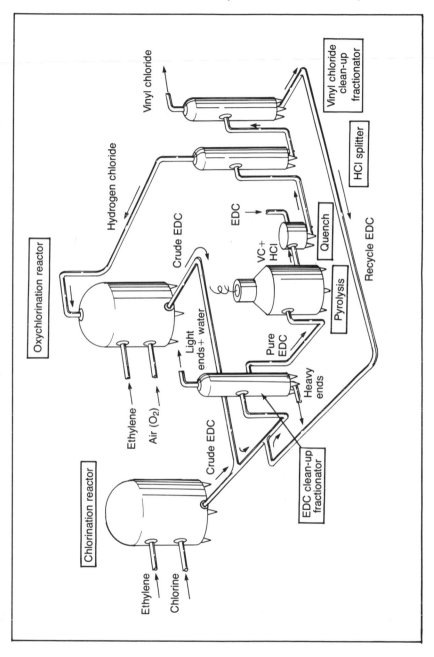

Fig. 9–2   Ethylene dichloride and vinyl chloride plant

Other Processes

The original process for vinyl chloride was the reaction of acetylene with hydrogen chloride in a fixed-bed reactor. The catalyst was activated carbon pellets impregnated with mercuric chloride. The next generation of VCM plants (built in the 1950s) looked much like the one in Fig. 9–2, except for the oxychlorination reactor. In those plants, the byproduct hydrogen chloride was reacted with acetylene, instead of ethylene and oxygen, in a process like the one just described. However, the chemical engineers were happy to replace the acetylene technology in the 1960s with the present state-of-the-art.

Another plant design involved cracking ethylene dichloride in the presence of caustic soda (sodium hydroxide or NaOH) to form vinyl chloride and sodium chloride (NaCl or table salt). The reaction was run at less severe (more cost effective) conditions, and the hardware was less expensive. However, the primary problem was what to do with all the salt. Mixed with water, it could be pumped into the ocean. Unfortunately, half the chlorine fed to the plant went with the salt water, and this was a financial waste.

Handling Characteristics

Sufficient evidence has proven that vinyl chloride can cause cancer of the liver after prolonged exposure to minute quantities (parts per million). Elaborate hardware precautions are taken to eliminate any VC release to the atmosphere. Personnel involved in producing or using vinyl chloride often wear respirators whenever the possibility of a leak exists.

Vinyl chloride vaporizes at about 7°F. Therefore, at normal temperatures it must be contained in high-pressure vessels to keep it liquid, including the tank cars and trucks used to transport it. These vehicles also must fly the hazardous material sticker en route. Like styrene, vinyl chloride is highly reactive and will polymerize with itself if it just sits in a tank. Phenol, in trace amounts, is an effective polymerization inhibitor and is normally added to vinyl chloride on the way to storage.

Ethylene dichloride is a much less nasty commodity. It does not need to be shipped in a pressurized vessel, but it is classified as a hazardous material and must be kept in a closed system.

Commercial Aspects

Most vinyl chloride is used to manufacture polyvinyl chloride in a lengthy process. Ethylene dichloride is usually used to manufacture vinyl chloride. Other uses of EDC are as a lead scavenger to clean gasoline; as an industrial degreaser and dry cleaning agent, perchloroethylene; and as an anesthesia, methyl chloroform. Both EDC and VC are traded commercially as a 99% pure grade. Vinyl chloride is usually designated as *inhibited,* which indicates the presence of phenol.

X.

ETHYLENE OXIDE AND ETHYLENE GLYCOL

From out of the past come the thundering hoofs
of the giant horse Silver

> from *The Lone Ranger Rides Again,*
> Fran Striker

All of the petrochemicals previously discussed and their immediate derivatives have a double-bonded pair of carbons imbedded in their structures. This characteristic makes the chemicals very reactive, which is why they are so useful as building blocks. In contrast, ethylene oxide (EO) has no double bond. Instead it has a three-member *heterocyclic* ring (*hetero* meaning one of the atoms isn't carbon, but oxygen). This cyclic oxide is often called an *epoxide.* (The suffix *ep* is from the Latin meaning "on" or "beside." In chemistry, it generally refers to the heterocyclic ring. The other common chemical with this suffix is epichlorohydrin.)

When ethylene oxide is formed, single bonds from two adjacent carbons are connected to an oxygen atom. A three-member ring is always in a strained condition, due to the geometry of the molecules. Because of the propensity of epoxides to relieve the strain, they are very reactive.

The key feature of ethylene glycol (EG) is the hydroxyl (—OH) groups, one on each of the two carbon atoms. The hydroxyls make ethylene glycol readily soluble in water. This behavior, plus EG's low freezing point, gives it the necessary properties for its most important end use, antifreeze. Also since ethylene glycol has such an affinity for water, it is used as a de-icer. When sprayed on ice, it combines with the water crystals and lowers the freezing point. This causes the mixture to melt and effectively keeps it in the liquid state.

Ethylene Oxide

Until the 1970s, the route to ethylene oxide was *ethylene chlorohydrin* in a two-step process. Ethylene, chlorine, and water were reacted

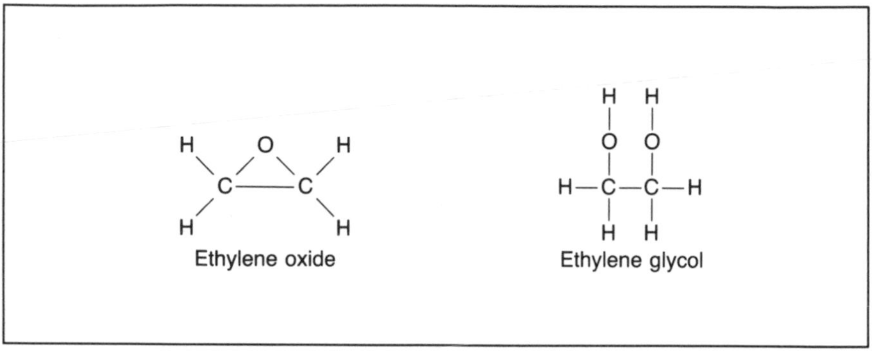

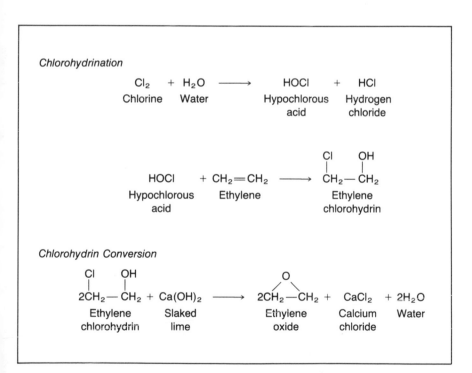

Fig. 10–1

Fig. 10–2   The obsolete chlorohydrin route to ethylene oxide

to form the ethylene chlorohydrin. Actually the chlorine reacted first with the water to make hypochlorous acid (HOCl) and hydrochloric acid (HCl). The hypochlorous acid then reacted with the ethylene molecules to form ethylene chlorohydrin, as shown in Fig. 10–2. In the second step, the chlorohydrin was exposed to hydrated lime, $Ca(OH)_2$. The lime picked off the chlorine and the hydroxyl hydrogen (the hydrogen in the —OH group). Only oxygen was left to satisfy the valence requirements of the dehydrogenated, dechlorinated molecule, and this step was done by forming a cyclic structure.

The problem with the chlorohydrin process was not the EO yield but the operating costs, particularly the cost of chlorine. Most of the fresh chlorine introduced into the process becomes calcium chloride. This compound is virtually worthless, and it must be hauled away as a solid and is really a diposal problem.

Much research was funded in the 1950s and 60s to find a process for oxidizing ethylene directly to ethylene oxide without numerous byproducts. Finally, Union Carbide found the silver bullet that did the job—a catalyst made of Ag_2O, silver oxide. The relationship between this catalyst and the reaction is unique. Silver oxide is the only substance with sufficient activity and selectivity. (Activity relates to conversion, selectivity to yield. *See* the appendix.) Moreover, ethylene is the only olefin affected in this way. The others, such as propylene and butylene, tend to oxidize completely, forming carbon dioxide and water. When silver oxide is used with ethylene, only a small amount of the second reaction in Fig. 10–3 occurs.

The process was so successful commercially that by the 1970s, this new chemistry had replaced the old. Adding some encouragement to this technological turnover was the availability of the obsoleted chlorohydrin plants previously used for manufacturing propylene oxide.

Process and Hardware Descriptions

The new EO plants are as simple as any you will read about in this book. The feeds are mixed, reacted, and then split into recycle and finished product streams, as shown in Fig. 10–4. The reaction occurs in the vapor phase. Ethylene and oxygen in the form of air are mixed and passed through a fixed-catalyst bed reactor. In some plants pure oxygen is added to the air to reduce the concentration of nitrogen, the other component of air. However, the cost of the pure oxygen supplement is a trade-off against the improvement in yield.

Sometimes catalyst promoters like inorganic barium or calcium salts are added to the catalyst bed. The reaction is carried out at 520–550°F

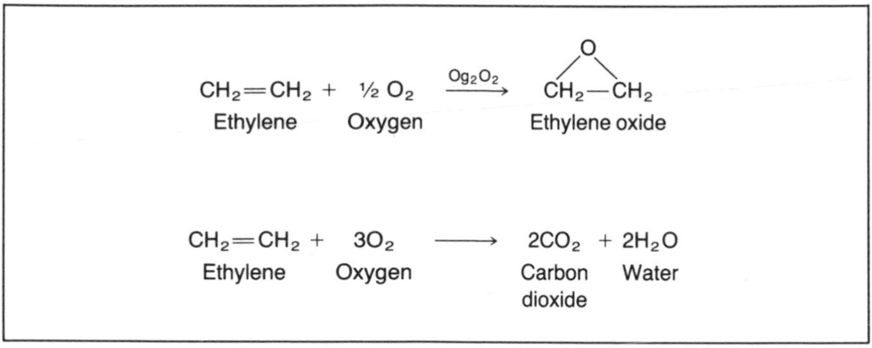

$$CH_2{=}CH_2 + \tfrac{1}{2} O_2 \xrightarrow{Og_2O_2} \underset{\text{Ethylene oxide}}{\overset{O}{CH_2{-}CH_2}}$$

$$\underset{\text{Ethylene}}{CH_2{=}CH_2} + \underset{\text{Oxygen}}{\tfrac{1}{2} O_2}$$

$$\underset{\text{Ethylene}}{CH_2{=}CH_2} + \underset{\text{Oxygen}}{3O_2} \longrightarrow \underset{\substack{\text{Carbon} \\ \text{dioxide}}}{2CO_2} + \underset{\text{Water}}{2H_2O}$$

Fig. 10–3 Direct oxidation of ethylene to ethylene oxide

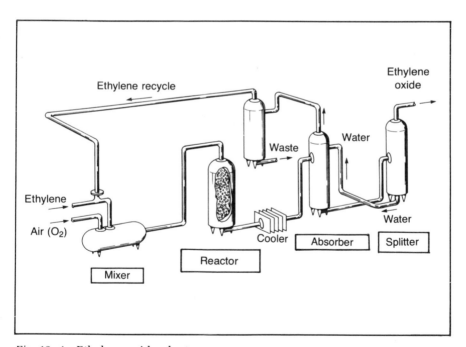

Fig. 10–4 Ethylene oxide plant

under slight pressure. The residence time of the feed in the reactor is only about one second, and the yield is about 60%.

Like most oxidations, this one is exothermic. The effluent from the reactor is cooled in a heat exchanger. A water-wash column separates the ethylene oxide, the byproducts, and the unreacted ethylene. This water wash is similar to the solvent recovery process described in chapter II. Ethylene oxide is absorbed by the water; the byproducts (mainly CO_2) and unreacted ethylene are not. The waste water and ethylene oxide are easily split in the next column.

When the byproducts and the ethylene are split, the ethylene is recycled to the reactor. (The byproduct splitter is not drawn correctly in Fig. 10–4 because some of the byproducts actually have lower boiling points than ethylene. Separation is really done in a series of columns.)

These reaction conditions are by no means standard from plant to plant. Process variations can include the following:

1. Catalyst concentration
2. Catalyst carrier
3. Type of catalyst bed—fixed or fluid
4. Use of catalyst promoter
5. Air-to-oxygen ratio
6. Operating conditions (temperature, pressure, and residence time)

Commercial Aspects

Uses. Ethylene oxide is an intermediary chemical and has multiple end uses. The predominant derivative is ethylene glycol, which uses up more than half the total produced. Biodegradable detergents are the second important outlet for EO. Others are fumigants (one of the oldest uses), latex paints, explosives, fibers, and pharmaceuticals.

Properties and Handling. Ethylene oxide is a colorless gas at room temperature, and it boils at 56°F. As a liquid, it is colorless, flammable, and very soluble in water and organic solvents. Ethylene oxide is traded commercially as a technical grade, 99.7% pure. Since it has a low boiling temperature, this compound must be shipped and stored in vessels that can withstand mild pressures. Also, trucks and tank cars must fly the hazardous material label.

Ethylene Glycol

The most widely used process for making ethylene glycol is shown in Fig. 10–5. This procedure is even simpler than the EO process. In

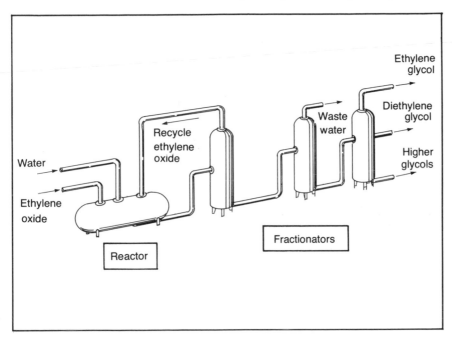

Fig. 10–5   Ethylene glycol plant

it, an ethylene-oxide ring is easily opened in the presence of water if a little sulfuric acid is added. Therefore, the processing scheme requires a mixing vessel for the following reaction to occur:

$$\underset{CH_2-CH_2}{\overset{O}{\triangle}} + H_2O \xrightarrow[\text{catalyst}]{\text{Acid}} \underset{CH_2-CH_2}{\overset{OH \quad\;\; OH}{|\qquad\;\; |}}$$

The other three pieces of hardware in Fig. 10–5 are used to recycle the ethylene oxide and to clean the ethylene glycol by splitting out the byproducts.

In the reactor, the operating conditions are slight pressure, 125–160°F, and a water-to-EO ratio of 1:10. The reaction is slow, so residence time in the reactor is about 30 minutes. This means the reactor must be large enough to accommodate high throughputs.

Newer, state-of-the-art technology has been commercialized in only a few plants. Two processes are evolving: *oxychlorination* and *acetoxyla-tion*. The first is very much like the old route to ethylene oxide via chloro-hydrin. The reaction occurs in the presence of diluted hydrochloric acid. Ethylene chlorohydrin forms in a transitory state because it is readily hydrolyzed (reacts with water) to form ethylene glycol at temperatures above 250°F. The catalyst to promote the reaction is thallic chloride ($TlCl_3$). The advantages of this process are the fewer stages involved (eliminating the EO step) and the HCl recovered for recycling.

Acetoxylation uses another exotic catalyst, tellurium oxide, with hydrobromic acid as a promoter. The reaction involves oxidizing ethylene in acetic acid at 320°F and 400 psi. Two reactions are involved, forming ethylene glycol mono-acetate and di-acetate. These products are subsequently hydrolyzed at 225–265°F to ethylene glycol and acetic acid.

Commercial Aspects

End Uses. About half of the ethylene glycol ends up as the principle ingredient in antifreeze. The remainder is primarily used to manufacture polyester fiber. To a lesser extent, ethylene glycol can make latex paints, surfactants, de-icing fluid for windshields and aircraft wings, and alkyl and polyester resins for films and coatings.

Properties and Handling. Ethylene glycol is a colorless, syrupy, sweet liquid. It is hydroscopic—absorbs water readily—and it lowers the freezing point of water. It is traded commercially as a high purity technical grade at 99% content. Ethylene glycol is a friendly liquid, and no particular precautions are needed to transport it by truck or tank car.

XI.

PROPYLENE OXIDE AND PROPYLENE GLYCOL

Something old, something new, something borrowed....

Anonymous, Wedding Rhyme

You have to talk about propylene oxide and glycol after ethylene oxide and glycol. It's not that the chemical configurations are so similar (they are) or that the process chemistry is about the same (it is). The fact is that most propylene oxide is now made in plants originally designed and constructed to produce ethylene oxide. As mentioned, the chlorohydrin route to ethylene oxide was abandoned in the 1970s in favor of direct oxidation. Producers found that these old plants provided the best technology and certainly the cheapest hardware to satisfy growing propylene-oxide demands.

Propylene Oxide

The chemical structure of propylene oxide differs from ethylene oxide by the methyl radical ($-CH_3$), as shown in Fig. 11–1. The difference is more than just a matter of geometric symmetry. The methyl radical affects the reactivity of the double bond in an adverse way. The reaction of propylene with chlorine or oxygen in the various manufacturing routes to PO is difficult to stop. Consequently, byproducts are an unavoidable nuisance, and the yields of propylene oxide are not as high as chemical engineers would like.

The Chlorohydrin Route

There are two major processes for the manufacturing propylene oxide, the old and the new. Prior to 1970, it was produced via the chlorohydrin process—a reaction of propylene with hypochlorous acid followed by dehydrochlorinating the propylene chlorohydrin with calcium hydroxide. That's a rather technical way of saying that a chlorine atom and an —OH radical are added to the propylene double bond. Then the

Ethylene oxide   Propylene oxide

Fig. 11–1

chlorine and the hydroxyl hydrogen are removed, leaving the oxygen bonded to two adjacent carbon atoms to form propylene oxide.

Four formulas describe the process. The first two are options for making the hypochlorous acid, using chlorine and water (or calcium hypochlorite and carbon dioxide if they're available and cheaper).

Hypochlorous acid

$$Cl_2 + H_2O \longrightarrow HOCl + HCl$$

or

$$Ca(OCl)_2 + CO_2 + H_2O \longrightarrow 2HOCl + CaCO_3$$

Propylene chlorohydrin formation

$$CH_3-CH=CH_2 + HOCl \longrightarrow CH_3-\overset{\overset{\displaystyle OH}{|}}{CH}-\overset{\overset{\displaystyle Cl}{|}}{CH_2}$$

Dehydrochlorinating to propylene oxide

$$2CH_3-\overset{\overset{\displaystyle OH}{|}}{CH}-\overset{\overset{\displaystyle Cl}{|}}{CH_2} + Ca(OH)_2 \longrightarrow$$

$$2CH_3-CH\overset{\displaystyle O}{\diagup\diagdown}CH_2 + CaCl_2 + 2H_2O$$

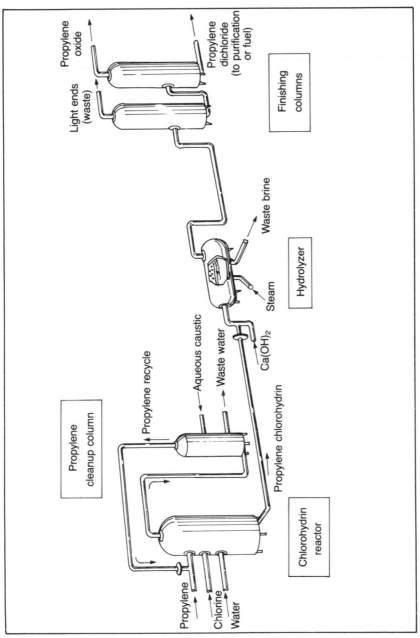

Fig. 11–2 Chlorohydrin route to propylene oxide

This process has the same disadvantages as the chlorohydrin route to ethylene oxide. Chlorine, via either route, is not always a cheap raw material. Furthermore, it's frustrating that the chlorine atoms don't even end up in the final product but are in calcium chloride, an almost worthless compound that usually presents a disposal problem. This route also has one disadvantage of its own—propylene dichloride formation. This byproduct is impossible to eliminate from the reaction. It can only be controlled. The demand for propylene dichloride is generally not sufficient to absorb the entire byproduct supply, so it too can bring a disposal problem. Luckily propylene dichloride burns, so it finds a home in the fuel system.

Finally, the hardware for this process is expensive because it uses corrosive compounds. If the EO plants had not been abandoned, more propylene oxide would be produced by alternative methods. However, the lure of the free hardware was a sufficient trade-off against the higher operating costs, and this route remains healthy, breathing heavily but healthy.

Research chemists successfully found an efficient, direct route to ethylene oxide, using silver oxide as a catalyst to oxidize ethylene. The search for the catalyst for propylene oxide has not been promising. Experimentation to date has turned in high-byproduct/low-PO yields.

The Chlorohydrin Hardware

Two reactions take place in the same vessel in the chlorohydrin plant. The formation of hypochlorous acid (HOCl) from chlorine and water and the reaction with propylene occur simultaneously in the reactor on the left in Fig. 11–2. Since propylene reacts readily with chlorine to form propylene dichloride, the hypochlorous (and hydrochloric) acid is kept diluted. Therefore, the concentration of propylene chlorohydrin leaving the reactor is only 3–5%. At higher concentrations, a separate phase or layer of propylene dichloride forms, which preferentially dissolves the chlorine and propylene and leads to a runaway propylene-dichloride yield.

The unreacted propylene is taken off the top of the reactor and is cleaned for recycling. By bubbling it through a diluted caustic solution (like sodium hydroxide, NaOH), the residual chlorine and hydrochloric acid are removed. Then the scrubbed propylene is ready as fresh feed or for use elsewhere in the plant.

The diluted propylene chlorohydrin stream is mixed with a solution of water and 10% slaked lime, $Ca(OH)_2$. This compound is pumped to a vessel called the hydrolyzer. The chlorohydrin rapidly converts to propylene oxide. The reaction is so fast that the PO must be sprung from the mixture before it continues on, forming propylene glycol. Steam is bubbled through the reactor, causing the propylene oxide to flash (vaporize) out of the reaction zone.

Vapor from the hydrolyzer contains water, propylene oxide, and whatever byproducts occur, including some chloropropenes,

$$CH_2\!=\!CH\!-\!CH_2Cl \text{ and } CH_2\!=\!\overset{\displaystyle Cl}{\overset{\displaystyle |}{C}}\!-\!CH_3 \text{; propionaldehyde, } CH_3CH_2CHO;$$

and of course, propylene dichloride, $CH_3\!-\!\overset{\displaystyle Cl}{\overset{\displaystyle |}{CH}}\!-\!\overset{\displaystyle Cl}{\overset{\displaystyle |}{CH_2}}$. Cleanup columns are used to segregate the PO into a technical grade.

Alternative Processes

Although direct propylene oxidation has not been successfully developed, several indirect oxidations have. These processes involve transferring oxygen from some other oxidized compounds. However, they are characterized and limited by side reactions, resulting in sizable byproduct formations.

For example, a few plants use isobutane and propylene to produce propylene oxide and either tertiary butyl alcohol (TBA) or isobutylene, both of which have a ready market. Other plants use ethylbenzene and propylene to produce PO and styrene, as discussed. In both processes, there is an intermediate reaction involving oxygen.

With isobutane, the first step forms tertiary butyl hydroperoxide (as shown in Fig. 11–3), by reacting isobutane with air at 200°F and 450 psi. In a second vessel, propylene then reacts with the hydroperoxide to form the two coproducts.

The advantages of these alternative processes are the high PO yield (about 80%), the coproducts, and the lower operating costs compared to the chlorine routes.

Commercial Aspects

Uses. About 60% of the propylene oxide produced makes polyurethane foams. You're probably sitting on a cushion stuffed with flexible polyurethane foam. Polyurethanes are the result of toluene di-isocyanate reacted with polyols or polyethers based on PO. Rigid polyurethane foams are used in load-bearing and structural applications.

The second largest application is fiberglass composites, which are reinforced polyesters. Fabricated products include boat hulls, shower

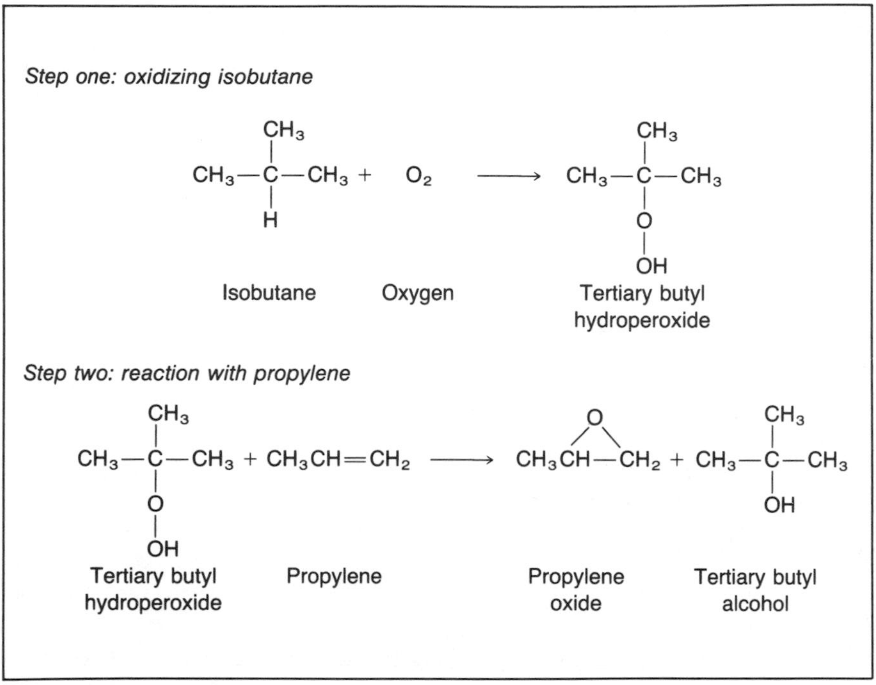

Step one: oxidizing isobutane

$$CH_3-\underset{\underset{H}{|}}{\overset{\overset{CH_3}{|}}{C}}-CH_3 \ + \ O_2 \ \longrightarrow \ CH_3-\underset{\underset{\underset{OH}{|}}{\overset{|}{O}}}{\overset{\overset{CH_3}{|}}{C}}-CH_3$$

Isobutane   Oxygen   Tertiary butyl
hydroperoxide

Step two: reaction with propylene

$$CH_3-\underset{\underset{\underset{OH}{|}}{\overset{|}{O}}}{\overset{\overset{CH_3}{|}}{C}}-CH_3 \ + \ CH_3CH{=}CH_2 \ \longrightarrow \ CH_3CH\overset{O}{\overbrace{\quad}}CH_2 \ + \ CH_3-\underset{\underset{OH}{|}}{\overset{\overset{CH_3}{|}}{C}}-CH_3$$

Tertiary butyl   Propylene   Propylene   Tertiary butyl
hydroperoxide   oxide   alcohol

Fig. 11–3 Propylene oxide-tertiary butyl alcohol from propylene/isobutane

stalls, furniture, and automotive parts. The polyester is generally reinforced with chopped glass fiber. Only about 20% of the propylene oxide goes to propylene-glycol and dipropylene-glycol production.

Properties and Handling

Propylene oxide is a low-boiling (93°F), flammable liquid that is readily soluble in both water and various organic solvents, such as alcohol, ether, and aliphatic and aromatic hydrocarbons. Commercial sales involve only technical grade (about 98%) and require a hazardous material shipping label. Standard bulk transport equipment (trucks, tank cars, and barges) can be used.

Propylene Glycol

This must be the quickest product treatment in this book—if you've read ethylene glycol. The process for propylene glycol is the same as for ethylene glycol. A little sulfuric acid in water opens the heterocyclic or epoxide ring, and the water provides the hydrogen and hydroxyl groups to form propylene glycol.

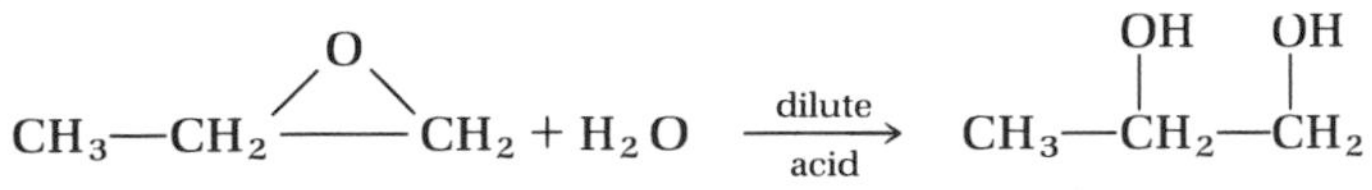

$$CH_3—CH_2 \overset{O}{\overbrace{\hspace{2em}}} CH_2 + H_2O \quad \xrightarrow[\text{acid}]{\text{dilute}} \quad CH_3—\overset{OH}{\underset{|}{CH_2}}—\overset{OH}{\underset{|}{CH_2}}$$

(Glycol is from the Greek root, *glyk*, meaning "sweet." The link is through the sugars, which have structures much like propylene glycol with multiple carbons and hydroxyl radicals.)

The hardware for propylene glycol is the same as that shown in Fig. 10–5. Just substitute propylene for ethylene to identify the streams. The byproduct stream is di-isopropylene glycol.

Commercial Aspects

Uses. Polyester resins use most of the propylene glycol manufactured. The balance goes to tobacco and cosmetic humectants (a chemical that retains moisture), automotive brake fluid, plasticizer for various resins, and a food additive. Di-isopropylene glycol has similar applications.

Properties and Handling. You can tell from the applications that propylene glycol is safe. It is nontoxic and nonflammable and is even fit for human consumption (in small doses). Propylene glycol is a colorless, odorless, sweet-tasting liquid, completely miscible or soluble in water. Propylene glycol is available in three grades: NF (99.99%), technical (99%), and industrial (95%).*

---

*NF = National Formulary

"'Is you deaf?' sez Brer Rabbit sezee. 'Kaze if you is, I can holler louder,' sezee."

The Wonderful Tar-Baby Story
Joel Chandler Harris (1848–1908)

Of the reactions introduced to produce the ten chemicals in the previous five chapters, four are induced by catalysts and two involve pyrolysis. The remainder require a little heat and pressure or an acid-based solution. Styrene and vinyl chloride are the products of pyrolysis. Cumene, ethylbenzene, ethylene dichloride, and ethylene oxide processes need special catalysts to make them work. The others depend on equipment design and operation. These reactions are summarized in Fig. 11–4.

The next six chapters cover a collection of petrochemicals not particularly related to each other. Synthesis gas is a basic building block. The alcohols, ketones, and acids are all close to each other on the petrochemical family tree. Maleic anhydride, acrylonitrile, and the acrylates? Well, they're all used to make polymers, and they had to be discussed somewhere.

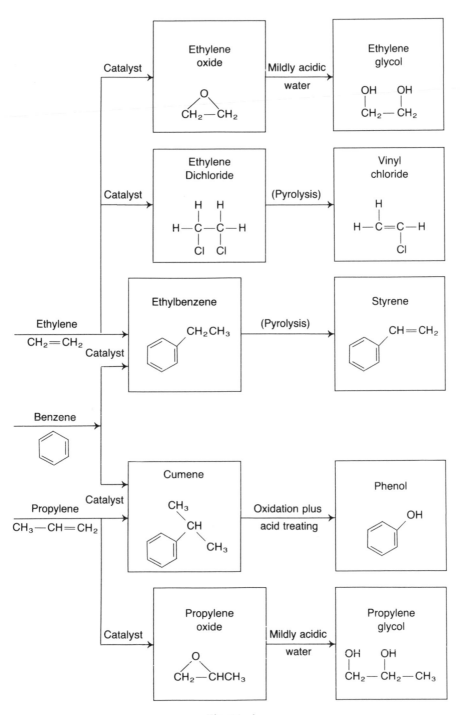

Fig. 11-4

XII.

METHANOL AND SYNTHESIS GAS

As the poet said, "Only God can make a tree"—probably because it's so hard to get the bark on.

Without Feathers,
Woody Allen (1935–)

The order in which you should approach these two subjects is reversed in the chapter title because you might be apprehensive if they weren't. After all, most people are familiar with methanol. It's methyl alcohol, CH_3OH, wood alcohol, carbinol, or if you're a student of medieval culture, *aqua vita*. But what is synthesis gas? It is not a common name because this chemical is usually not handled in commercial transactions.

The term synthesis gas refers to various mixtures of carbon monoxide (CO) and hydrogen (H_2) used to manufacture certain petrochemicals. In the early 19th century, it was produced by passing steam over coke at very high temperatures. Today synthesis gas is made primarily from natural gas (methane). Therefore, a few paragraphs about synthesis gas, how it's made, and how it can be used to synthesize other petrochemicals will be beneficial, especially since two very important chemicals, ammonia and methanol, are derived from it.

Synthesis Gas

Mother Nature hasn't provided any convenient sources of pure carbon monoxide and hydrogen. Some of each are contained in natural gas, but usually not in sufficient quantities to justify producing the gas. However, these two compounds, either in the combined state or separated, are readily convertible to a number of commercial compounds. With that as a motivator, several processes have been developed to convert natural gas to synthesis gas. Natural gas is primarily methane (CH_4), which provides a source of carbon and hydrogen. Air or water contribute the other necessary ingredient, oxygen.

Synthesis gas should not be confused with synthetic natural gas, SNG. Both are sometimes called syngas. However, SNG is basically methane made from petroleum products, like naphtha or propane, or made from coal. It's used as a substitute for or supplement to natural gas.

Synthesis Gas Processes

The two predominant methods to make synthesis gas are *steam reforming* and *partial oxidation*. Both are quite simple. Steam reforming involves passing methane or naphtha plus steam over a nickel catalyst. If methane is the feedstock, the reaction is as follows:

$$CH_4 + H_2O \longrightarrow CO + 3H_2$$

The reaction is endothermic (heat must be added). It also must be carried out in hardware much like a cracking furnace, with high temperatures and pressures. Undesirable reactions occur, resulting in the formation of carbon dioxide and carbon. Both are a nuisance, particularly carbon because it deactivates the catalyst.

The other method, partial-oxidation of methane, is represented as follows:

$$CH_4 + \tfrac{1}{2}O_2 \longrightarrow CO + 2H_2$$

Like the steam-reforming method, this process occurs at severe conditions (high temperatures and pressures) but with no catalyst. The reaction is called partial oxidation because it is kept from producing carbon dioxide by limiting the amount of oxygen fed to the process.

The partial-oxidation method is normally used for heavier feedstocks, everything from naphtha to residual fuel, in those places where natural gas or light hydrocarbons (ethane, propane, or butane) are not readily available. The yield of carbon monoxide is not 100% in either process. You can see in Table 12–1 that plenty of CO_2 also forms as a byproduct.

TABLE 12–1
Synthesis Gas Composition
(Typical yield, based on methane feed)

	H_2	CO	CO_2
Steam reforming	75	15	10
Partial oxidation	50	45	5

To tailor synthesis gas to the various applications shown in Table 12–2, the composition can be modified in several ways. To reduce CO content, steam is mixed with the synthesis gas and is passed over a catalyst, where the following reaction occurs:

$$CO + H_2O \longrightarrow CO_2 + H_2$$

The carbon dioxide can be removed by solvent extraction. In this reaction, the CO content decreases, and fresh hydrogen is added to the synthesis-gas mixture.

If it's necessary to increase the carbon-monoxide content in the synthesis gas, a little carbon dioxide can be added to the steam reforming plant feed (methane). Then the following reactions also occurs in the reactor:

$$CH_4 + 3CO_2 \longrightarrow 4CO + 2H_2O$$

Most of the water generated reacts with more methane to produce hydrogen and carbon monoxide:

$$CH_4 + H_2O \longrightarrow CO + 3H_2$$

All of the combinations of synthesis gas shown in Table 12–2 can be achieved with the above reactions, except the ammonia plant feed.

TABLE 12–2
Synthesis Gas Applications

Mixtures	Main uses
Pure H_2	Refinery hydrotreating and hydrocracking
$3H_2:1N_2$	Ammonia plant feed (Haber process)
$1H_2:1CO$	Aldehydes and alcohols (Oxo reactions)
CO or $CO + H_2O$	Acids (formic and acetic)
$2H_2:1CO$	Methanol plant feed
$2H_2:1CO$	Olefins (Fischer-Tropsch reaction)
$3H_2:1CO$	Hydrocarbons (Fischer-Tropsch reaction)

Ammonia

The Haber process is the basic route to ammonia (NH_3). A mixture of nitrogen and hydrogen are passed over an iron-oxide catalyst promoted with aluminum oxide at 700°F and 4,000 psi.

$$N_2 + 3H_2 \longrightarrow 2NH_3$$

Why does ammonia develop from synthesis gas? And where does the nitrogen come from? Of course, synthesis gas provides the hydrogen. Also if the synthesis-gas process is partial oxidation, an air-separation plant probably is associated with it. This unit separates the oxygen from the nitrogen to make the synthesis gas and leaves the nitrogen for feed to the ammonia plant.

In most ammonia plants, facilities to remove carbon monoxide from the feed are usually included because CO can poison the catalyst. The carbon-monoxide reaction with water to produce carbon dioxide and hydrogen is usually used, and the carbon dioxide is removed by solvent extraction. (Typical solvents are ethanolamine or an aqueous solution of potassium carbonate.)

The Plant

The steam-reforming route to synthesis gas is shown on the left side of Fig. 12–1. The primary reaction occurs in the furnace tubes, which are packed with a nickel-oxide catalyst. The reactor temperatures are about 1,600°F; pressure is approximately 500 psi.

To get the proper $CO:H_2$ ratio for the methanol plant, carbon dioxide is added to the reactor feed. The CO_2 is recovered from a slipstream of furnace stack gas. (When hydrocarbons are burned as fuel, the main combustion products are carbon dioxide and water.) The slipstream in Fig. 12–1 goes to a column for purifying the carbon dioxide, which separates small amounts of carbon monoxide and nitrous oxides from the CO_2. In practice, the CO_2 purification facilities are more elaborate, since an absorption process is generally used.

A final step before sending the synthesis gas to the methanol plant is to cool the gas. Cooling is usually done in a series of heat exchangers to recover the energy.

Commercial Aspects

Most synthesis gas produced is captive. That is, it's consumed by the manufacturer. Normally synthesis-gas plants are integrated into the adjacent application plant. When a two-party transaction is involved, the properties of the synthesis-gas stream usually are specified in a contract. There are no standards that apply to this stream.

The only practical way to move synthesis gas is by pipeline. Even in two-party transactions, the pipelines are usually no longer than a mile or two. Beyond that, the pipeline capital cost starts to affect the economics of the applications.

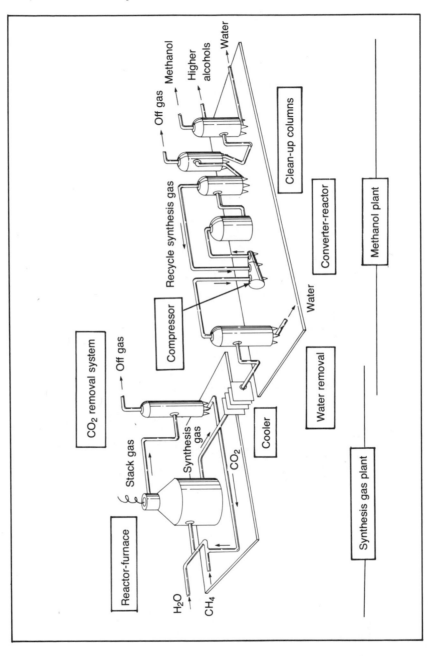

Fig. 12–1 Synthesis gas and methanol plants

Methanol

There's a good reason why methanol is commonly called wood alcohol. The early commercial source was the destructive distillation of fresh-cut lumber from hardwood trees. When wood is heated without access to air at temperatures above 500°F, it decomposes into charcoal and a volatile fraction. One of the liquids that can be condensed is methanol. Hence, methanol was named wood alcohol or wood spirits.

Since 1923, methanol has been made commercially from synthesis gas. This route provides most of the methanol today. The chemical formula for the reaction is simple:

$$CO + 2H_2 \rightleftharpoons CH_3OH$$

The reaction is run at about 750°F and 4,000 psi over a copper or zinc/chromium-oxide catalyst. The double arrows indicate that the reaction can go in either direction. However, the CH_3OH requires only one-third of the volume that the CO and the $2H_2$ need. At the high pressures involved, the reaction is pushed to the right since CH_3OH takes up less room.

Originally synthesis gas was processed to remove all of the carbon dioxide. Now methanol catalysts have been developed to handle both carbon oxides. Therefore, this reaction also takes place in most methanol plants:

$$CO_2 + 3H_2 \longrightarrow CH_3OH + H_2O$$

Since some hydrogen ends up as water, the reaction is considered less efficient. Therefore, carbon monoxide is the preferred route.

The Plant

All of the processes for synthesizing methanol involve these basic steps:
1. Steam reforming natural gas and adding carbon dioxide to adjust the $CO:H_2$ ratio to 1:2;
2. Compression at 4,000–5,000 psi;
3. Synthesis in a catalytic converter;
4. Purification and distillation.

The hardware needed is shown in Fig. 12–1. To protect the compressors, a water knock-out column in front is necessary. This column keeps water slugs from forming during compression, which can send turbine blades flying around the plant.

The compressed gas is heated and passed through a reactor that has baskets of catalyst. In between the baskets are heat exchangers. The reaction is exothermic; however, it is sensitive to the temperature. Therefore, heat must be removed rapidly.

The effluent from the reactor contains only 5–20% methanol because the one-pass conversion is very low. After the cooling and pressure let-down phases, the liquid methanol can be removed and further purified by distillation. The unreacted synthesis gas is recycled to the reactor.

Methanol of 99% purity is obtained. Byproducts are 1–2% dimethyl ether (CH_3OCH_3), about 0.5% higher alcohols (ethyl, propyl, isobutyl, and higher), and some water.

Commercial Aspects

Properties and Handling. Methanol is a colorless, volatile liquid at room temperature with an alcoholic smell. It mixes with water in all proportions and burns with a pale blue flame. Methanol is highly toxic. As little as one-fifth of a shot (10cc) can cause blindness. Larger amounts kill. Methanol should never be applied to the body as a rubbing alcohol because the vapors are very toxic.

Methanol boils at 148°F, so it can be transported in closed systems by conventional tank trucks and cars, ships, and barges. The red hazardous materials markings are required.

Uses. About half of the methanol produced is converted to *formaldehyde*. That's not because the embalming business is so good. Formaldehyde is a feedstock for many products: phenolic resins, which are used as adhesives in plywood; hexamethylene tetramine, used in electronic plastics; and pentaerythritol, used to make enamel coatings, floor polish, and inks. In the textile business, formaldehyde also is used to make fire retardants, mildew resistant linens, and permanent press clothing. About 20% of the methanol produced is used to make dimethyl terephthalate (for polyester fibers) and methyl methacrylate (for plastics). Another application of methanol is the methyl chloride in silicone rubber, including the caulking and sealing compounds that set at room temperatures.

Methanol is also used to denature ethanol. There's not much difference between synthetic ethyl alcohol and real ethanol made from rye and other grain. However, to protect the economic interests of the grain growers, the law requires that as much as 10% methanol be added to the synthetic product. Ethanol denatured in this way is very toxic and can cause headaches, dizziness, vomiting, blindness, and comas, depending on how much is consumed. That effectively eliminates the use of synthetic ethyl alcohol for human consumption.

XIII.

THE OTHER ALCOHOLS

'Twas a woman that drove me to drink, but I never had the courtesy to go back and thank her.

W. C. Fields (1879–1946)

There are many other commercial alcohols besides methanol. This chapter treats those that are traded in the largest volumes: ethyl alcohol (EA), isopropyl alcohol (IPA), normal butyl alcohol (NBA), and 2-ethyl hexanol (2-EH). The higher alcohols (octyl-, decyl-, dodecyl-, and tridecyl-alcohol) also are discussed.

You may want to review the section on alcohols in chapter I. There you'll find that organic alcohols have two distinct parts: the hydrocarbon chain or ring (R in shorthand) and the hydroxl group (—OH), which together form the alcohol signature R—OH. Often, but not always, the alcohol is named after whatever the R is: CH_3—CH_2—OH is ethyl alcohol; CH_2=CH—OH is vinyl alcohol. However, C_6H_5OH, a benzene ring with a hydroxyl radical, is phenol (Fig. 13–1).

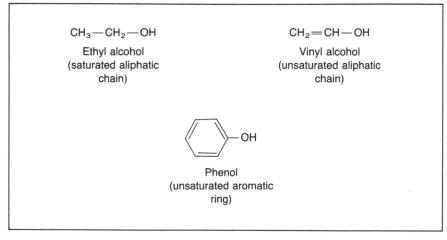

Fig. 13–1   Alcohols

When there is more than one hydroxyl group per molecule, it is still an alcohol. Ethylene glycol, CH_2—CH_2 is a *polyhydric* alcohol, as

$$\quad\quad\quad\quad\quad\quad\quad\quad\quad\quad\quad\quad OH\quad OH$$

is glycerine, CH_2—CH—CH_2.

$$\quad\quad\quad\quad OH\quad OH\quad OH$$

Like Caesar's Gaul, petrochemical processes used for manufacturing alcohols today are divided into three parts.

1. Direct hydration. An olefin is reacted with water:

$$CH_2{=}CH_2 + H_2O \longrightarrow CH_3{-}CH_2{-}OH$$

$$\quad\text{Ethylene}\quad\quad\quad\quad\quad\quad\text{Ethyl alcohol}$$

2. Indirect hydration. An olefin is reacted with an inorganic acid, and the resulting inorganic ester is reacted with water:

$$CH_3CH{=}CH_2 + H_2SO_4 \longrightarrow CH_3{-}\underset{\underset{\textstyle OSO_3H}{|}}{CH}{-}CH_3$$

$$\quad\text{Propylene}\quad\quad\quad\quad\text{Isopropyl hydrogen sulfate}$$

$$\underbrace{\quad\quad\quad\quad\quad\quad\quad\quad}_{A}$$

$$A + H_2O \longrightarrow CH_3{-}\underset{\underset{\textstyle OH}{|}}{CH}{-}CH_3 + H_2SO_4$$

$$\text{Isopropyl alcohol}$$

3. Oxo reaction. An olefin reacted with synthesis gas produces an aldehyde, which is hydrogenated (hydrogen is added) to produce alcohol:

$$CH_3{-}CH{=}CH_2 + CO + H_2 \longrightarrow$$

$$\quad\text{Propylene}\quad\quad\text{Synthesis}$$
$$\quad\quad\quad\quad\quad\quad\text{gas}$$

$$CH_3{-}CH_2{-}CH_2{-}CHO + CH_3{-}\underset{\underset{\textstyle CH_3}{|}}{CH}{-}CHO$$

$$\text{Normal butyraldehyde}\quad\quad\text{Isobutyraldehyde}$$

$$\underbrace{\quad\quad\quad\quad\quad\quad\quad\quad\quad\quad\quad\quad}_{A}$$

$$A + H_2 \longrightarrow CH_3-CH_2-CH_2-CH_2-OH$$

Normal butyl alcohol

$$+ \; CH_3-\underset{\underset{\textstyle CH_3}{|}}{CH}-CH_2-OH$$

Isobutyl alcohol

Ethyl Alcohol

Sugar fermentation in the presence of yeast to produce ethyl alcohol in the form of wine goes back beyond recorded history. Originally the sugar came from grapes. Later starch from grain, potatoes or corn was used. The yeast came from living matter in the form of mold or fungus. This particular yeast contains the enzyme *zymase*. This enzyme catalyzes the sugar fermentation. Sugar (in grape juice) mixed with yeast reacts slowly—weeks, months, maybe years—to form ethyl alcohol and carbon dioxide, as well as minor amounts of some aldehydes. Some of the nonalcoholic contents also can be separated by distilling.

Alcoholic beverages in the United States must be made exclusively by the fermentation process, not the petrochemical process. This has nothing to do with the chemistry involved. It's a law enacted to protect the grain growers, not the consumers. The convention for identifying the alcoholic content of beverages is *proof.* An alcohol that is 100 proof is 50% alcohol; 86 proof scotch is 43 percent alcohol, and so on. The proof divided by two is the percentage of alcohol. Pure alcohol is 200 proof.

Until World War I, fermentation accounted for all of the ethyl alcohol produced in the U.S. In 1919, a petrochemical route based on ethylene and sulfuric acid was developed commercially. By 1935, only 10% of the ethyl alcohol was produced this way. With the rapid improvements of ethylene technology, the share quickly grew to 90% by the 1960s, when the direct hydration method became popular. It virtually replaced all indirect hydration processes by the 1970s. The advantages of the direct method were higher yields, less pollution, and lower plant maintenance due to less corrosion. All of these led to better economics. Currently, about 95% of domestic ethyl alcohol is produced via the direct catalytic hydration of ethylene.

$$CH_2{=}CH_2 + H_2O \xrightarrow{\; H_3PO_4 \;} CH_3-CH_2-OH$$

The Process

The chemical reaction for ethyl alcohol occurs in a single reactor, as shown in Fig. 13–2. The remaining facilities are handling and cleanup hardware.

Ethylene is compressed to 1,000 psi before it is mixed with water and heated to 600°F. The two reactants, both in a vapor phase, are fed down a reactor filled with a catalyst. The catalyst is phosphoric acid on a porous, inert support (usually diatomaceous earth or silica gel).

The ethylene-to-ethyl-alcohol conversion per pass through the reactor is only 4–6%, so most of the ethylene must be recycled. However, the reactor effluent is first cooled and caustic washed—sodium hydroxide is added to remove the phosphoric acid. As the effluent cools, the ethyl alcohol liquefies and the ethylene can easily be separated. Then the ethylene is scrubbed by sloshing it through water prior to recycling.

The mixture from the bottom of the separator and the scrubber is crude ethyl alcohol. That is, it contains ethyl alcohol, water, and all byproducts. Further distillation produces a conventional EA-water mixture that boils at a single temperature (an azeotrope).

To purify ethyl alcohol highly, the EA-water azeotrope is distilled with a third liquid such as benzene. The benzene forms a ternary (three-compound) azeotrope with ethyl alcohol and water, boiling at a constant temperature of 157.5°F. It also forms a binary azeotrope with ethyl alcohol, which boils at 162.5°F. These two azeotropes are removed by distillation, leaving pure ethyl alcohol that otherwise would boil at 173.1°F. Some of the byproducts often recovered are diethyl ether (an anesthetic) and acetaldehyde, which can be easily hydrogenated to ethyl alcohol.

Commercial Aspects

Uses. Ethyl alcohol has been used as a chemical intermediate to manufacture acetaldehyde and acetic acid. This application is fading because recent process improvements have been made to oxidize ethylene directly. (Over 80% of today's acetaldehyde is made via the direct-oxidation route.) Ethyl alcohol remains in demand as a solvent in personal care products (such as after-shave lotion and mouthwash), cosmetics, pharmaceuticals, and even surface coatings. It is a precursor for ethyl chloride, which is used to make tetraethyl lead, diethyl ether (the anesthetic), and ethyl acetate, a solvent for coatings and plastics. (The scent of nail polish remover is ethyl acetate.)

Properties and Handling. Ethyl alcohol is a colorless, flammable liquid (good for flambé) with a characteristic odor that is recognized

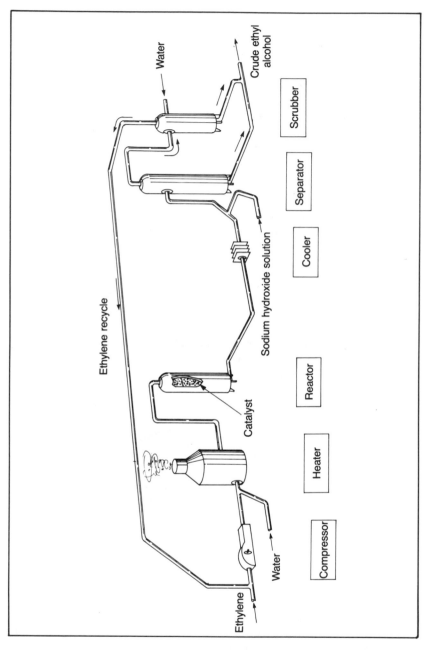

Fig. 13–2 Ethyl alcohol plant

universally. This alcohol is soluble in water (and club soda) in all proportions. It is commercially available as 190 proof (the 95% EA-water
azeotrope) and as absolute (200 proof). Frequently ethyl alcohol is
denatured to avoid the high tax associated with 190- and 200-proof
grades. Methanol and sometimes formaldehyde are common dena-
turants used to prevent consumption as an alcoholic beverage.

Because it is flammable, ethyl alcohol is transported as a hazardous
material in tank cars and trucks, barges, tankers, and drums.

Isopropyl Alcohol

Technically, isopropyl alcohol can be made via the direct hydration
method, but the severe operating conditions make this route energy-
intensive. Therefore, the prefered process for IPA is indirect hydration.
This method is more economical than direct hydration because lower
pressures and temperatures are involved. (Since ethylene does not
react at the lower temperatures, indirect hydration is impractical for
ethyl alcohol.)

The specifications for the feed to an IPA plant can be loose.
Refinery-grade propylene, even with some small amounts of ethane and
ethylene, can be used because the C_2's and propane don't react. They
just pass through the process. In fact, the process acts as a kind of C_3
splitter, since about 50% of the propylene is reacted out to isopropyl
alcohol in each pass through the reactor.

The Process

Propylene is absorbed by concentrated sulfuric acid to form isopropyl hydrogen sulfate. Subsequently this compound is hydrolyzed
with water to isopropyl alcohol and diluted sulfuric acid. The propylene
stream is fed into the bottom of a column (Fig. 13–3) that is packed with
baffles to give intimate contact. Concentrated (85%) sulfuric acid is introduced at the top of the vessel.

As the acid and propylene slosh past each other, about 50% of the
propylene reacts with the sulfuric acid to form sulfate. The reaction is
exothermic, so the tower contents must be cooled continually to maintain a 70–80°F temperature. This cooling minimizes byproducts, particularly propylene polymers. Higher olefins in the propylene feed
(usually C_4 and C_5) are absorbed by the sulfuric acid, forming sulfates

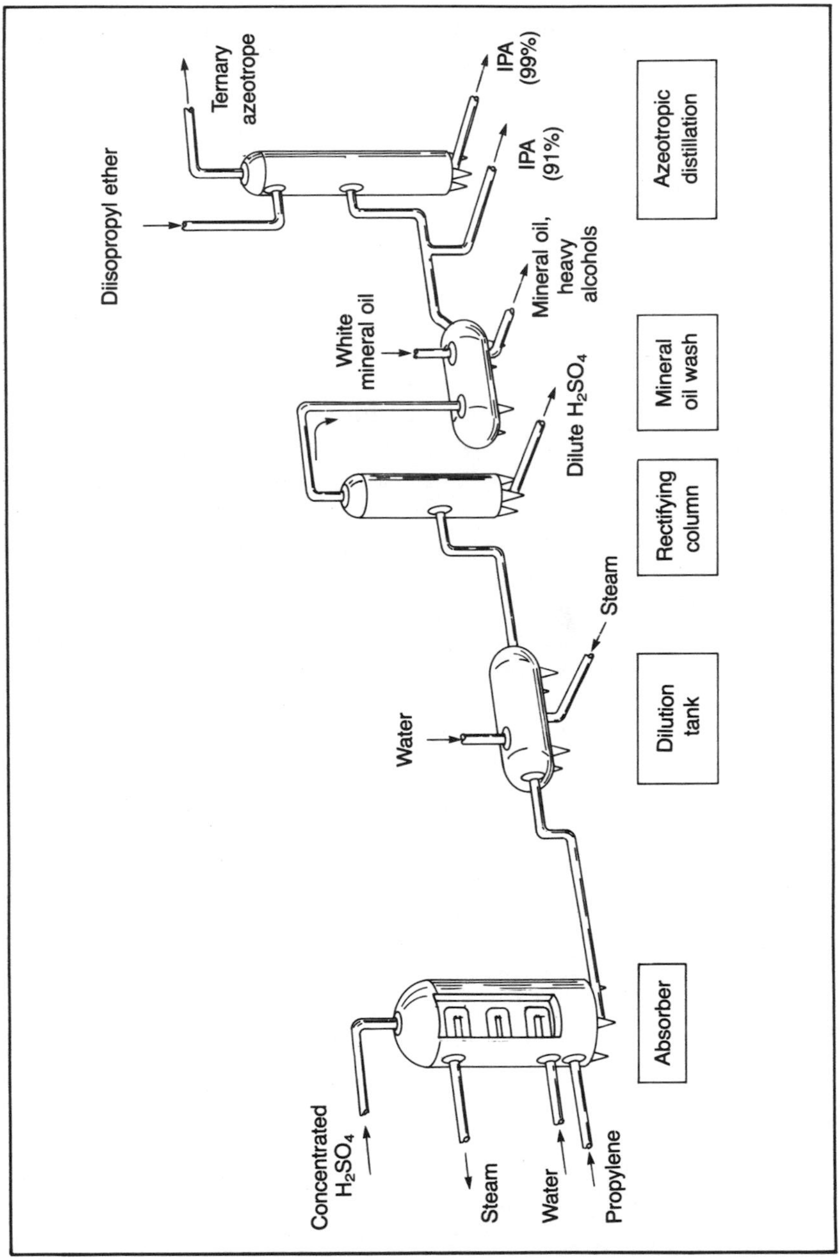

Fig. 13–3 Isopropyl alcohol plant

and bisulfates. These byproducts must be removed in the cleanup facilities. The yield of isopropyl alcohol from propylene is about 70%.

Propylene from the reactor top can be recycled to the feed, although the concentration may need to be boosted by a splitter or by adding some chemical-grade propylene. The effluent from the reactor bottom is dumped into a lead-lined tank and is diluted with water and steam, cutting the unreacted sulfuric acid to about 20%. Mixing sulfuric acid and water is exothermic, and that heat plus a little steam is sufficient to hydrolyze the isopropyl hydrogen sulfate to IPA. With a little more steam the crude isopropyl alcohol flashes (vaporizes) from the dilution tank to a rectifying column for concentration. The diluted hydrogen-sulfate stream is cleaned and reconstituted to a higher concentration for reuse.

At the rectifying column, a 91% IPA/water azeotrope distills from the top, carrying most of the other organics. This azeotrope is washed with mineral oil to remove the C_4's, C_5's, and higher alcohols. It is further treated with sodium hypochlorite to give the water-white technical grade, which is still the 91% IPA/water azeotrope.

Like ethyl alcohol, the absolute (99+%) grade is made by forming a ternary azeotrope. In this case, DIPE (diisopropyl ether) is used. The azeotrope boils off at 142.5°F in a 91.1/5.8/3.1% weight combination of DIPE-IPA-water. The high purity isopropyl alcohol, which boils at 180.5°F, leaves the column bottom.

You may wonder why the process produces isopropyl alcohol instead of (normal) propyl alcohol. With the exception of ethylene, direct or indirect hydration of an aliphatic olefin always produces an alcohol with the hydroxyl radical preferentially attached to the double-bonded carbon having the fewest hydrogen atoms.

Commercial Aspects

Uses. About 50% of the isopropyl alcohol produced is used to make acetone (dimethyl ketone). Fortunately, the 91% IPA-water azeotrope is an adequate feedstock. Like ethyl alcohol, IPA is a chemical intermediate (for isopropyl esters and hydrogen peroxide) and a solvent (for surface coatings, personal care products, and cosmetics).

At one time isopropyl alcohol was used as a gasoline additive to prevent cold weather stalling, but it has been displaced primarily by DIPE. And of course, IPA is used as rubbing alcohol because it has an innocuous nontoxic odor, a low boiling (vaporization) temperature, and a moderate heat of vaporization. Isopropyl alcohol dries rapidly but won't give you frostbite like liquid butane can.

Properties and Handling. Isopropyl alcohol is a colorless, flammable liquid with that characteristic, rubbing-alcohol odor. It's soluble in water in all proportions, as well as in most organic solvents. It is commercially available in technical (91%), chemical (98%), and absolute (99+%) grades. Shipments by rail, truck, and drum, are routine. However, since it is flammable, hazardous materials warnings are required.

Normal Butyl Alcohol and 2-Ethyl Alcohol

There's another type of nomenclature that's applied to alcohols (among other things): *normal, secondary,* and *tertiary.* For example, with C_4 alcohols, the hydroxyl group can be connected to one of the following:

1. A primary carbon atom (attached to only one other carbon atom);
2. A secondary carbon atom (attached to two other carbon atoms);
3. A tertiary carbon atom (attached to three other carbon atoms).

Therefore, these chemical arrangements are possible:

$CH_3—CH_2—CH_2—CH_2—OH$ Normal butyl alcohol, NBA, 1° alcohol, or $R—CH_2—OH$

$CH_3—CH_2—\underset{\underset{OH}{|}}{CH}—CH_3$ Secondary butyl alcohol, SBA, 2° alcohol, or $R—\underset{\underset{R}{|}}{CH}—OH$

$CH_3—\underset{\underset{CH_3}{|}}{\overset{\overset{CH_3}{|}}{C}}—OH$ Tertiary butyl alcohol, TBA, 3° alcohol, or $R—\underset{\underset{R}{|}}{\overset{\overset{R}{|}}{C}}—OH$

Normal butyl alcohol was first recovered in the 1920s as a byproduct of acetone manufactured via cornstarch fermentation. That route is almost extinct. About 25% is now made from acetaldehyde, $CH_3—CHO$, which is reacted with itself to form aldol, $CH_3—CHOH—CH_2—CHO$. Aldol is then dehydrated and hydrogenated to yield normal butyl alcohol. The remaining NBA production is based on the Oxo process.

The Oxo process is used in a number of applications to extend the length of an olefin chain by one carbon. The reaction is between an olefin and synthesis gas (carbon monoxide and hydrogen) in the presence of a cobalt catalyst. It produces a mixture of aldehydes (the —CHO signature group), which readily undergo hydrogenation to alcohols. One important feature of the process is that it produces only *primary* alcohols.

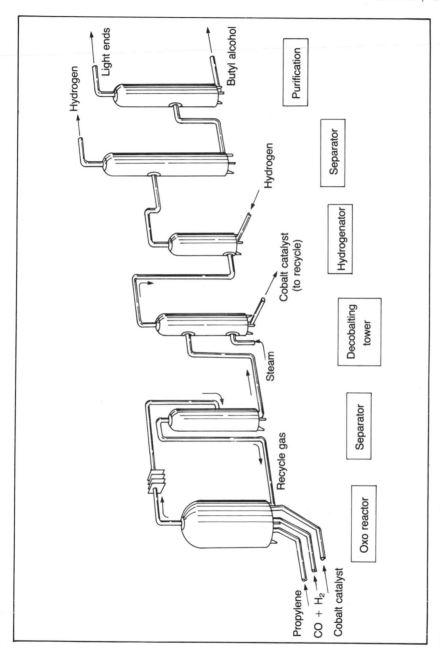

Fig. 13–4 Oxo process for normal butyl alcohol

Most other petrochemical processes yielding alcohols produce *secondary* or *tertiary* alcohols.

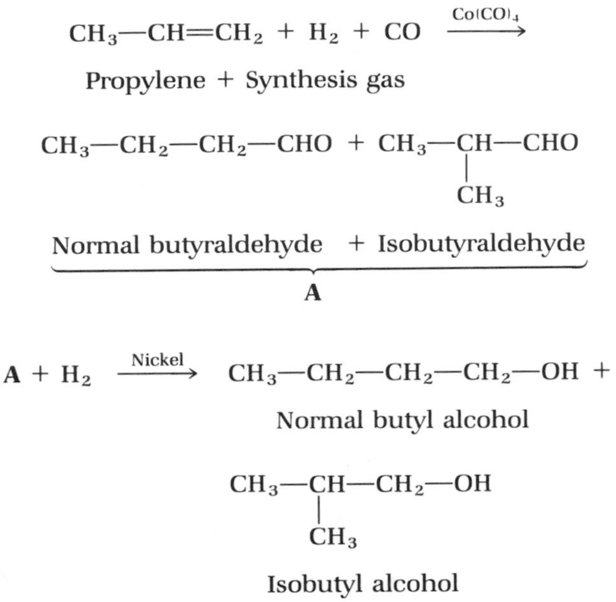

$$CH_3—CH{=}CH_2 + H_2 + CO \xrightarrow{Co(CO)_4}$$

Propylene + Synthesis gas

$$CH_3—CH_2—CH_2—CHO + CH_3—\underset{\underset{CH_3}{|}}{CH}—CHO$$

Normal butyraldehyde + Isobutyraldehyde

A

$$A + H_2 \xrightarrow{Nickel} CH_3—CH_2—CH_2—CH_2—OH +$$

Normal butyl alcohol

$$CH_3—\underset{\underset{CH_3}{|}}{CH}—CH_2—OH$$

Isobutyl alcohol

Both the butyl aldehydes and alcohols are formed in this process, but the demand for normal butyl alcohol greatly exceeds the iso-isomer. By using mild operating conditions, the ratio of normal to iso increases. However, the total yield goes down. Recycling the isobutyral-dehyde to the feed also fools the reaction and keeps down the iso yield. Most processes yield an overall ratio of normal to iso of about 4:1.

One of the oldest *high molecular-weight aliphatic* alcohols is 2-EH (2-ethyl hexyl alcohol or 2-ethyl hexanol). What does it have in common with NBA? Both are made from propylene via the Oxo process, and both have the same aldehyde intermediate—normal butyraldehyde.

The aldehyde undergoes aldol condensation. It dimerizes to form a hydroxyaldehyde. (Dimer has the same Latin root, *-meros*, as isomer, monomer, and polymer, and means "part." A dimer is a chemical union of two molecules of the same compound.) The resulting C_8 dimer is also an aldehyde that can be hydrogenated to give 2-EH.

Dimerization:

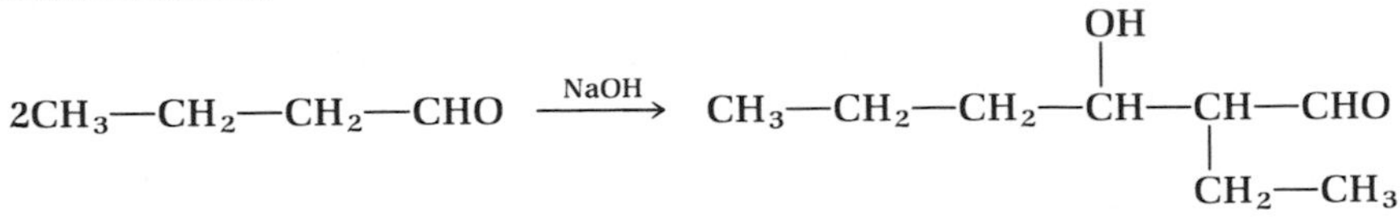

$$2CH_3—CH_2—CH_2—CHO \xrightarrow{NaOH} CH_3—CH_2—CH_2—\underset{\underset{CH_2—CH_3}{|}}{\overset{\overset{OH}{|}}{CH}}—CH—CHO$$

A

Hydrogenation:

$$A + 2H_2 \xrightarrow{\text{Nickel}} CH_3-CH_2-CH_2-CH_2-\underset{\underset{CH_3}{\overset{|}{CH_2}}}{CH}-CH_2-OH + H_2O$$

2-ethyl hexanol

The name 2-EH becomes apparent from the layout of the molecule. The straight C_6 chain gives the *hexyl*; the ethyl side chain, $-CH_2-CH_3$, gives the *ethyl*. The ethyl group is connected to the second carbon from the right end of the chain bearing the hydroxyl group, giving the *two*.

Commercial Aspects

Uses. The motivation for first recovering normal butyl alcohol in the 1920s was its use as a lacquer solvent. That application is even stronger today. The NBA vapors from lacquer drying are nontoxic and virtually nonflammable. The other uses for normal butyl alcohol are as plasticizers and chemical intermediates, mostly for esters.

The most important use for 2-ethyl hexanol is in manufacturing *di-2-ethyl hexyl phthalate*, which is used as a plasticizer that makes polyvinyl chloride flexible. About 75% of the 2-EH goes to this application. The growth of 2-EH acrylate is becoming another important application. This acrylate is an effective construction adhesive and can be found in some surface coatings. Other uses for 2-EH include industrial solvents, dispersing and wetting agents, and chemical intermediates.

Properties and Handling. NBA and 2-EH are nonvolatile, colorless, nontoxic liquids, with relatively high boiling points of 243°F and 364°F, respectively. Normal butyl alcohol is only slightly soluble in water, and 2-EH is insoluble. A rule of thumb may be helpful: like dissolves like. Methanol is very much like water because the hydroxyl group in both molecules (CH_3OH and HOH) is a significant part of the molecule. The same principle is true for ethyl and isopropyl alcohol. However, in normal butyl alcohol and 2-ethyl hexanol, the hydroxyl is minor and the carbon chain is significant. The analogy between compounds and other solvents, including organics usually holds true—like dissolves like.

Both NBA and 2-EH are available in technical grade (98–99%) and are transported in normal equipment. No hazardous material label is required.

NBA Process

The feeds to an NBA plant are chemical-grade propylene, synthesis gas, and an oil-soluble cobalt salt (dissolved in the propylene), cobalt naphthenate. These chemicals are fed to a reactor maintained at 250–300°F and 3,500–4,000 psi. The reactor effluent contains unreacted gases, a catalyst, and the aldehydes. The pressure is reduced in a separator, and the unreacted gases are recycled. The hydrogenation step takes place in the conventional way, in a vessel packed with a catalyst, and the aldehydes and hydrogen are admixed at 200–300°F and 600–1,200 psi. The catalyst is usually nickel or copper chromite on an inert carrier, such as kieselguhr, silica gel, or alumina. The crude butyl alcohols are then separated and are purified by distillation.

2-EH Process

The same NBA front-end process is used to make the butyraldehyde feed for 2-EH. However, the butyraldehyde is dimerized (sometimes called *aldol condensation* when done in the presence of an alkaline catalyst). Some plants even combine the Oxo process and the aldol process (then referred to as the Aldox process).

In the Aldox route, the butyraldehydes produced in the oxo step are fed with hydrogen to a reactor filled with a nickel catalyst, maintained at 300°F at 2,500 psi. Distilling the effluent gives 2-EH yields of 93% with isobutyl alcohol as a byproduct.

At one time, the isobutyraldehyde had to be split out before the dimerization reaction, but catalysis improvements have permitted cogenerating both C_4 alcohols and 2-ethyl hexanol.

SBA and TBA Processes

Secondary and tertiary butyl alcohols are produced by the absorption of butylene and isobutylene in concentrated sulfuric acid. The processes are similar to the indirect hydrolysis route for isopropyl alcohol.

Secondary butyl alcohol is a colorless, high-boiling-point (212°F) liquid with a pleasant odor. The tertiary alcohol is a white solid (melting point is 78°F) with a camphor-like odor. Both alcohols are traded as technical grade (99% purity) and need a hazardous (corrosive) materials label.

Higher Alcohols

The two major routes to higher alcohols (C_8 and larger) are the Oxo process and the Alfol® process, which was developed by Conoco. As discussed, the Oxo process transforms an olefin to an aldehyde to an alcohol. Three higher alcohols produced via the Oxo process in commercial volumes are iso-octyl (from heptene, C_7H_{14}), decyl alcohol (from nonene, C_9H_{18}), and tridecyl alcohol (from dodecene, $C_{12}H_{24}$). Yields run 70–75%.

In the Alfol® process, even-number-carbon, linear, primary alcohols are produced. The process involves the reaction of aluminum, hydrogen, and ethylene under controlled polymerization. It's not as complex as it may sound. Aluminum, hydrogen, and ethylene are reacted to make triethyl aluminum, an aluminum atom with three individually attached ethyl radicals. The next step is *oligomerization.** That is, each of the ethyl radicals polymerizes with additional ethylene molecules, forming *trialkyl* aluminum compounds. (The *alkyl* is the ethyl radical that has grown 2, 4, 6, or more carbon atoms by adding one, two, three, or more ethylene molecules. The operating conditions and catalyst determine the chain length.)

The trialkyl aluminum is oxidized with air. Then it is hydrolyzed in water or diluted sulfuric acid to yield a crude-alcohol mixture and aluminum hydroxide (alum). After separation, the crude alcohols are washed and fractionated.

The process produces only *primary linear* alcohols having an even number of carbon atoms. Commercial alcohols produced by this method are octyl, decyl, dodecyl, tetradecyl, and hexadecyl alcohols.

Uses. The higher alcohols are important raw materials for producing high molecular-weight esters (phthalates, adipates, sulfates), which are used as plasticizers or are converted to anionic detergents. Further reactions with ethylene oxide yield nonionic detergents, which are used as light-duty liquid dishwashing soaps, heavy-duty powdered laundry soaps, and bar soaps. Other minor applications are as lacquer solvents, synthetic lubricants, perfumes, antifoaming agents, herbicides, and lube-oil additives.

---

*From *olig,* meaning "several" and *meros,* meaning "parts."

XIV.

ACETONE, METHYL ETHYL KETONE, AND METHYL ISOBUTYL KETONE

Names are not always what they seem. The common Welsh name Bzjxxllwcp is pronounced Jackson.

Mark Twain (1835–1910)

There's no need for ill-humored comments about the ketones being a trio of pop singers. They're a family of organic compounds that all have the ketone signature, . Somewhere in the middle of a hydrocarbon chain, the two hydrogens attached to a carbon are replaced by a double-bonded oxygen. Ketones come in many sizes and shapes. Convention for naming them is to refer to the radicals (alkyl groups) attached to the ketone signature. In Fig. 14–1, the three commercially traded ketones with the largest volumes are shown: acetone, MEK, and MIBK.

Acetone

Right up front you need to know that acetone and dimethyl ketone (DMK) are the same chemically. Acetone has been used extensively as a solvent since pre-World War I. The commercial routes to producing acetone included the destructive distillation of wood and the fermentation of either starch or corn syrup. The development of olefin technology permitted the more efficient petrochemical methods to replace the original processes by the late 1930s.

One convenient way to classify the present processes for making acetone is to separate them into two categories: byproduct and on-purpose. You'll recall that acetone is one of the outturns of the cumene-to-phenol process described in chapter VII. Acetone from that process falls into the category of byproduct production because the rate at which acetone is produced does not depend solely on anticipated acetone demand. Often the demand for phenol dictates the rates at which the

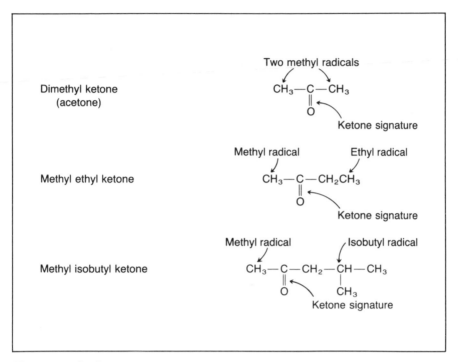

Fig. 14–1 The ketones

phenol plant is run. Therefore, the acetone supply can present a disposal problem, or a shortage can exist.

The swing supply of acetone comes from plants that produce acetone on-purpose by dehydrogenation or oxidation of isopropyl alcohol (IPA) and by direct oxidation of propylene. Almost as much capacity is available to produce on-purpose acetone as there is via the phenol route.

Process Description

Two on-purpose routes to acetone, dehydrogenation and oxidation, have plants that look very similar. As in other petrochemical processes, the part of the plant where the chemical reactions take place is simple. It's all the mechanical processes, such as heating, cooling, and especially separating, that fill up the plant site.

The dehydrogenation process is shown in Fig. 14–2. In this plant, the isopropyl-alcohol feed is heated to approximately 900°F in a preheater and is charged to a reactor at about 40–50 psi pressure. The reactor is filled with a catalyst of zinc oxide deposited on pumice. Pumice

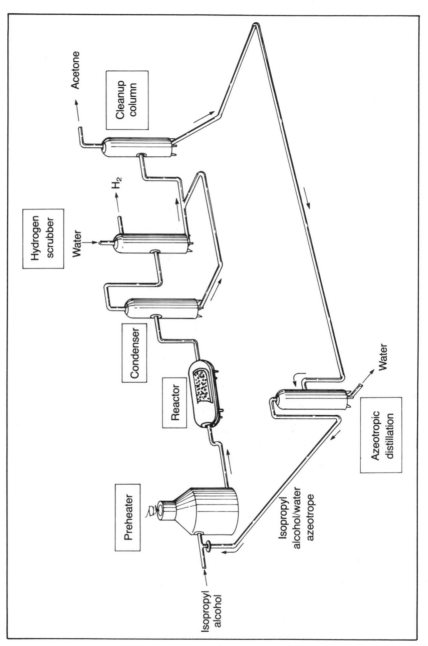

Fig. 14–2 IPA-to-acetone plant

is a fine powder of silica dioxide, which is glass. It has many fine pores in which the catalyst can reside. Therefore, it has a very large surface area exposed to the isopropyl alcohol. Hydrogens pop off the middle carbon and the OH group, forcing the double bond to the oxygen (the ketone signature).

The effluent from the reactor is acetone, unreacted IPA, and hydrogen. In one of the several columns that follow the reactor (called a *hydrogen scrubber*), water is used to split out the hydrogen. Both isopropyl alcohol and acetone are highly soluble in water, but hydrogen is not. Therefore, by washing the effluent with water, the hydrogen bubbles out the top and the IPA/acetone and water come out the bottom. The process is just like the solvent-recovery scheme explained in chapter II.

Acetone can be separated from the unreacted IPA and water in a fractionator. However, if the water in the IPA recycle stream requires a special treatment, IPA and water will form an azeotropic solution (like the one discussed in chapter III). Together these two compounds boil off in a 90/10 mixture at a temperature different than the boiling points of either. Consequently, the stream recycled to the reactor from the azeotropic distillation column contains about 10% water. (The IPA dehydrogenation step that follows is not affected. The water is just a little extra volume to heat, cool, and pump around.) The bottoms from the azeotropic distillation are almost pure water and are recycled to the water-wash column.

In the oxidation plant, air is added to the feed stream. The reactor contains a silver- or copper-oxide catalyst, and the reaction temperature is 750–1,100°F. Water is formed instead of hydrogen (see Fig. 14–2). However, the water-wash column is still needed to split out the nitrogen, which was introduced with the air. Both processes yield about 85–90% acetone.

The improved acetone technology is the Wacker process, a direct oxidation route. Had you been in charge of building the acetone business from scratch, you probably would not have built an IPA-to-acetone plant if you'd known about the Wacker process. This system catalytically oxidizes propylene at 200–250°F and 125–200 psi over palladium chloride with a cupric (copper) chloride promoter. The yields are 92–94%. The hardware for this process is probably less expensive than that for the combined IPA/acetone plants. However, once those plants were built, the economies of the Wacker process were not sufficient to shut down existing systems and start all over. Therefore, growth in the Wacker technology has only picked up since 1970, and some of this progress has been shared with new cumene-to-phenol/acetone plants.

There are several other methods that are of minor importance. Acetone is a coproduct in the reaction between isopropyl alcohol and *acrolein* for producing *allyl alcohol*. It also is a coproduct when IPA is oxidized yielding acetone and H_2O_2, *hydrogen peroxide*—the principal ingredient of bleach.

Commercial Aspects

Uses.   Acetone is used in two basically different ways, as a chemical intermediate and as a solvent. As an intermediate, acetone produces MIBK, *methyl methacrylate* (used to make plexiglass products), *bisphenol A* (a raw material for epoxy and polycarbonate resins), and higher molecular-weight glycols and alcohols.

As a solvent, acetone is used in varnish, lacquer, cellulose acetate fiber, cellulose nitrate (an explosive), and as a carrier solvent for acetylene in cylinders. Acetylene is stored at about 225 psi but is so explosively reactive that as an extra precaution the cylinder is filled with asbestos wool soaked in acetone. Acetylene is extremely soluble in acetone, and the asbestos keeps it from sloshing around when the cylinder is half empty. Acetone also is used in smaller volumes for manufacturing pharmaceuticals and chloroform (the anesthetic).

Properties and Handling.   Acetone is a mobile, colorless, volatile, highly flammable liquid. It has an odor that makes you think you're in a hospital. Acetone dissolves in water, alcohol, ether, and most other organic solvents. That's why it's usually included in paint-brush cleaner. It dissolves almost any paint base and then can be washed away with water.

Acetone is sold commercially in three grades, U.S.P. (99%), C.P. (99.5%), and technical (99.5%).* It is shipped in run of the mill tank trucks and cars and in drums. The hazardous material shipping placard must be displayed for this highly flammable liquid.

Methyl Ethyl Ketone

Most of what you've read in the previous section about acetone also applies to methyl ethyl ketone (MEK). The processes for making MEK can be categorized broadly into byproduct and on-purpose; the more popular processes start with larger molecules. The three primary manufacturing methods to MEK are as follows:

---

*U.S. Pure and Chemical Pure.

1. Dehydrogenating secondary butyl alcohol (instead of isopropyl alcohol);
2. Oxidizing butylene (instead of propylene) with air;
3. Catalytically oxidizing butane to form acetic acid and the by-product MEK.

Not only is there a similarity in the chemistry of the dehydrogenation route, the hardware is almost identical to the plant shown in Fig. 10–2.

$$CH_3-\overset{\displaystyle OH}{\underset{\displaystyle |}{CH}}-CH_2-CH_3 \longrightarrow CH_3-\overset{\displaystyle O}{\overset{\displaystyle \|}{C}}-CH_2-CH_3 + H_2$$

SBA   MEK

The heated secondary butyl alcohol (SBA) vapors pass over a zinc-oxide catalyst at 750–1,000°F and at atmospheric pressure. The separation of the reactor effluent into MEK, water, hydrogen, and recycle SBA is about the same. The overall yields are approximately 85–90%.

There is a more energy-efficient version of this process that takes place in the liquid phase. A catalyst of very fine Raney nickel or copper chromate, carried in a heavy, high-boiling-point solvent, is mixed with SBA. At 300–325°F, the SBA undergoes dehydrogenation to MEK. As it does, the MEK and hydrogen immediately vaporize, leaving the reaction medium in gaseous form. Then the molecules need only to be separated from each other.

Another on-purpose route to MEK is to oxidize butylene directly, called the Wacker process:

$$2CH_2{=}CH-CH_2-CH_3 + O_2 \longrightarrow 2CH_3-\overset{\displaystyle O}{\overset{\displaystyle \|}{C}}-CH_2-CH_3$$

Butylene   MEK

With reaction conditions of 200–225°F, 150–225 psi and a palladium chloride-cupric chloride catalyst, MEK yields are 80–90%. The operating costs of the Wacker process for MEK (as well as acetone and several other petrochemicals) are relatively low. However, the plant is built of expensive materials. Since the catalyst solution is corrosive, critical vessels and piping must be titanium based and the reactor is made of rubber-lined, acid-resistant brick.

The byproduct route to MEK accounts for only a modest portion of the total supply. Plants designed to produce acetic acid by directly oxidizing butane can be run to produce almost no MEK. However, optimum operating costs balanced against market product prices usually warrant shifting to a 60/40 acetic acid/MEK outturn.

Commercial Aspects

Methyl ethyl ketone is used in various ways as a solvent. It owes part of its popularity to the fact that it is considered nontoxic, and is not an air pollutant. As a result, MEK can be used as a solvent in coatings, such as vinyl, nitrocellulose, and acrylic. In these applications, MEK flashes (vaporizes or dries) quickly at room temperatures, leaving the coating behind. Since the MEK ends up in the atmosphere, its non-polluting character is important. This ketone also is used as the solvent in lube-oil dewaxing, wood pulping, and toluene extraction and in manufacturing ink.

The physical characteristics of MEK are those of acetone. It is colorless, mobile, flammable, and sweet smelling (if that's what you call a hospital smell). It's very soluble in water and in most common organic solvents. There are only two grades commercially traded: technical (99%) and C.P. (99.95%). Shipping and handling procedures are similar to acetone.

Methyl Isobutyl Ketone

The process for methyl isobutyl ketone (MIBK) is more complicated than the one-step conversion method for acetone and MEK. The three-step process shown in Fig. 14–3 manufactures MIBK, starting with acetone. First, the acetone is condensed with itself. That is, it passes over a catalyst, and two acetone molecules chemically react to form diacetone alcohol. Both the ketone $\left(\begin{matrix} O \\ \| \\ -C- \end{matrix}\right)$ and alcohol (—OH) signatures are attached to diacetone alcohol. The catalyst is an alkaline compound like $Ca(OH)_2$ (calcium hydroxide or soda lime), and the reaction is run at about 32°F.

In the second step, the diacetone alcohol is dehydrated and the —OH group and a hydrogen atom are clipped off to produce mesityl oxide. Dehydration is done by mixing the diacetone alcohol with a sulfuric-acid catalyst at 212–250°F. In the third step, the mesityl oxide is hydrogenated (hydrogen is added) to MIBK by heating it to the vapor stage at 300–400°F, mixing it with hydrogen, and passing it over a copper or nickel catalyst at 50–150 psi.

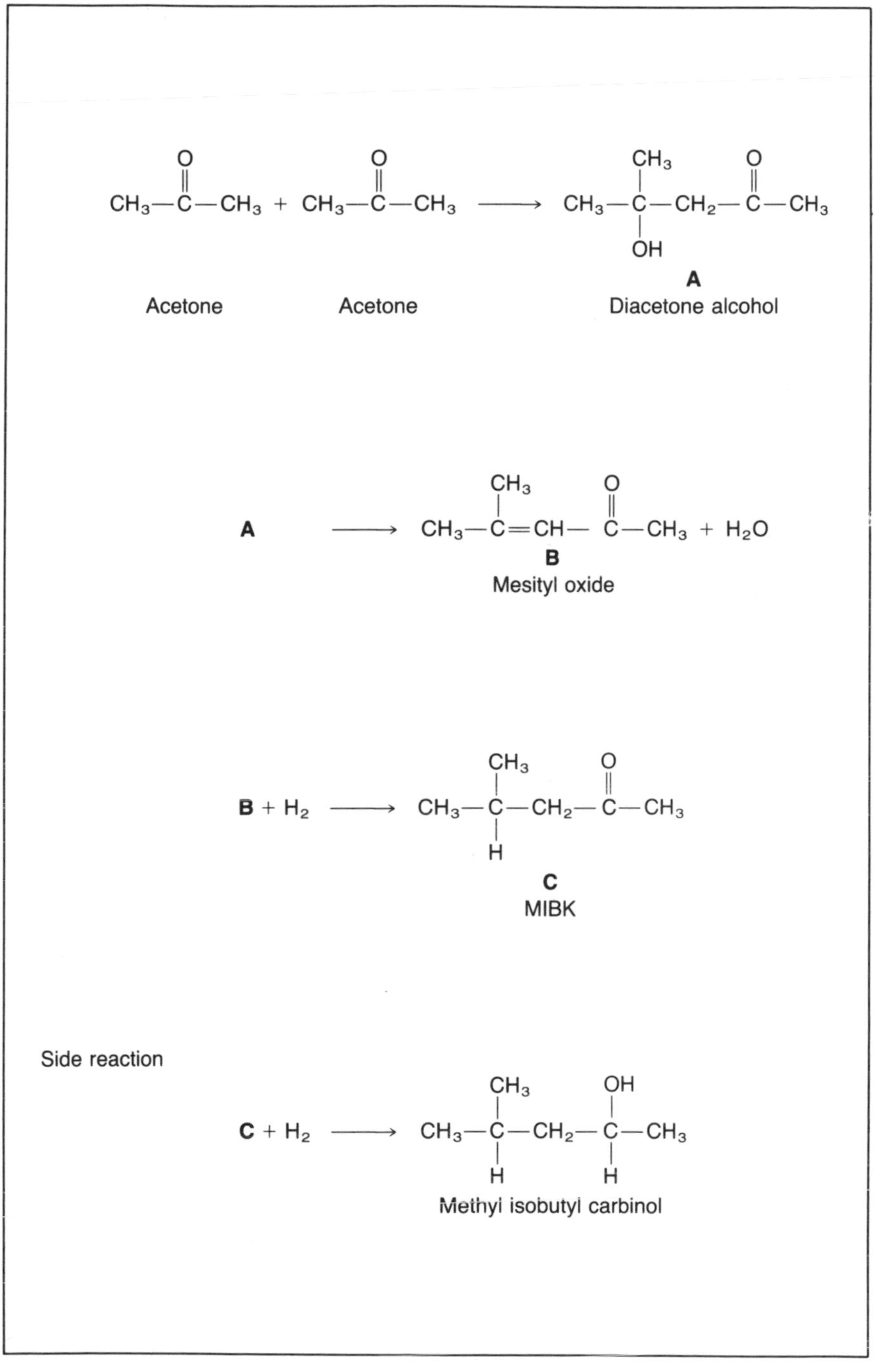

Fig. 14–3 MIBK plant reactions

One problem with this process is the difficulty in controlling the final step. As the MIBK forms, it also tends to hydrogenate further to methyl isobutyl carbinol. Adding hydrogen wipes out the ketone signature, replacing it with the hydroxyl group, —OH. This unavoidable byproduct, methyl isobutyl carbinol, is separated from the MIBK in a fractionator downstream.

Alternative Processes

A small amount of MIBK is made from a new European-originated process. A complex catalyst system involving palladium metal and a cation-exchange resin is used. The reaction permits production directly from acetone to MIBK.

A method that's similar in concept involves going directly from isopropyl alcohol to a mixture of acetone and MIBK. The process is confidential, and details are not available.

Commercial Aspects

Uses. The applications of MIBK read a lot like those of MEK. In the 1960s and 70s, MIBK rapidly replaced ethyl acetate and butyl acetate as a solvent in resins. However, MEK is now a better competitor in many areas because MIBK is alleged to be a pollutant. Some unique applications for MIBK include metallurgical extraction (particularly uranium), a reaction solvent in pharmaceuticals, and an adhesive. If you stretch the definition of application, it also can be used to manufacture methyl isobutyl carbinol.

Properties and Handling. MIBK is a colorless liquid and has a pleasant, almost fruity, odor. Unlike acetone and MEK, it is only slightly soluble in water. Most commercial trade in MIBK is in the technical grade (98.5%). Bulk shipments of MIBK can be handled in conventional tank trucks and cars, but the hazardous material markings must be displayed.

THE ACIDS

Eye of a newt and toe of a frog,
Wool of a bat and tongue of a dog.

MacBeth, Shakespeare (1564–1616)

Dozens of organic acids are used in petrochemicals processing, but there are four that account for 80% of the volume: acetic, acrylic, adipic, and phthalic acids. These compounds have little in common with each other besides the carboxyl signature grouping, written as —COOH and drawn as C—OH. The group is so called because it's a com-

$$\overset{\|}{O}$$

bination of the *carbo*nyl (—C=O) and hyd*rox*yl (—OH) groups.

Acids can be aliphatic or aromatic and mono-, di-, or polycarboxylic. Aliphatic acids are often referred to as the fatty acids because many of them were originally obtained by hydrolyzing animal fat or vegetable oil. (Aliphatic is from the Greek word *aleiphatos,* meaning "fat.") Like other aliphatics, the fatty acids are straight chain organic compounds with some branching. Aromatic acids, on the other hand, have the benzene ring directly connected to the carboxyl signature group. *Dicarboxylic* acids have carboxyl signature groups attached in two places. *Monocarboxylics* have only one, and of course, the *poly* acids have three or more.

Acetic Acid

Acetic acid is the simplest and largest member of the aliphatic acid family. It's a methyl group attached to the acid signature group, CH_3COOH. There are several places you're likely to find acetic acid or its derivatives. Acetic acid is the natural component of vinegar that gives it the characteristic smell. (*Acetum* is the Latin word for vinegar.) Acetic also is used to make acetates, which are polymerized and processed into coatings and adhesives (polyvinyl acetate) and fibers (cellulose acetate, like Arnel).*

Manufacturing Acetic Acid

Acetic acid is a byproduct of natural wood-alcohol (methanol) production, but very little of the process remains. At one time, it accounted for all of the commercial acetic-acid production. However, most of it is now made by oxidizing butane or acetaldehyde, which is made from ethyl alcohol.

$$CH_3-CH_2-CH_2-CH_3 + O_2 \longrightarrow CH_3-\underset{\underset{O}{\|}}{C}-OH + Byproducts$$

Butane   Acetic acid

$$CH_3-CHO + \tfrac{1}{2}O_2 \longrightarrow CH_3-\underset{\underset{O}{\|}}{C}-OH$$

Acetaldehyde   Acetic acid

Acetic acid looks a lot like methanol. It just has an extra carbon and oxygen. You'd think there might be a fit there. Sure enough, several new plants based on a methanol/carbon-monoxide reaction have been built. This process may be much cheaper than the other two routes and may be the preferred method from now on.

$$CH_3-OH + CO \longrightarrow CH_3-\underset{\underset{O}{\|}}{C}-OH$$

Methanol   Carbon   Acetic acid
            monoxide

Acetates are acetic esters. Remember from chapter I that an ester has the signature group $-\underset{\underset{O}{\|}}{C}-O-R$, and the ester's name usually ends in the suffix *ate*. The acetate group is an ester made from acetic acid by replacing the carboxyl hydrogen with an R group: $CH_3-\underset{\underset{O}{\|}}{C}-O-R$. For example, if R is the vinyl group $-CH{=}CH_2$, you have vinyl acetate, $CH_3-\underset{\underset{O}{\|}}{C}-O-CH{=}CH_2$.

The Acids / 141

There's also a process based on light naphtha, which is pretty much like the butane operation. Table 15–1 compares the vital statistics of each route.

Acetic Acid Plants

Acetaldehyde Processes. A stainless-steel, water-jacketed kettle is charged with concentrated (99%) acetaldehyde and a catalyst. Then air is bubbled through it for about 12 hours. The gases from the top of the kettle are bubbled through water to scrub them and are discharged to the atmosphere. Some acetaldehyde ends up in the scrubbing water, which is distilled and recycled. The mixture in the kettle, crude acetic acid, is distilled to 99% purity. Acetic acid of this grade is called *glacial* acetic acid.

Vapor- and liquid-phase processes also are available, which combine the acetaldehyde-from-ethanol and acetic-acid-from-acetaldehyde reactions into one plant. Ethanol and acetaldehyde are oxidized over a catalyst at the same time. The produced acetaldehyde is recycled, so the only fresh feed is really ethanol.

TABLE 15–1
Acetic Acid Process Conditions

Process	Catalyst	Reaction conditions, °F	psi	Yield, %	Byproducts
Butane oxidation	Cobalt acetate	300–450	800	57	Acetaldehyde, acetone, methanol
Light naphtha oxidation	Manganese acetate	400	750	40	Formic acid, propionic acid, acetone
Acetaldehyde oxidation	Manganese acetate	150	0	95	None
Methanol carbonylation (BASF)	Cobalt iodide	400–475	4,000	87	None
Methanol carbonylation (Monsanto)	Rhodium iodide	350–475	400	99	None

Butane and Light Naphtha Process. The butane and light naphtha process is run under high pressure (750–800 psi), so the butane or light naphtha (C_4's and C_5's) remain liquid. However, other light gaseous products vaporize and escape. The butane/naphtha is added to a vessel that contains acetic acid and cobalt acetate. Air is bubbled through and is stirred or mixed vigorously. The volatile byproducts come out the top: methane, carbon dioxide and monoxide, low molecular-weight alcohols, ketones, acids, and unreacted air. Crude acetic acid is drawn from the bottom of the vessel and is distilled to give glacial acetic acid in about 50% yields.

Methanol. BASF introduced technology in 1960 to make acetic acid from methanol and carbon monoxide instead of from ethylene. Monsanto subsequently improved the catalysts, permitting operations at lower pressures. The methanol and carbon monoxide, of course, come from a synthesis gas plant. The process proceeds at about 350°F and 400 psi with a rhodium-iodide catalyst and a sodium-iodide promoter to produce acetic-acid yields of 99%. Byproducts include only small amounts of dimethyl ether and methyl acetate. The high yields, conditions that are not too severe, and bypassing the capital-intensive ethylene plant make this route attractive.

Commercial Aspects

Applications. About 75% of the acetic acid ends up in vinyl and cellulose acetates. Polyvinyl acetate is used for plastics, coatings and paints. Cellulose acetate is predominantly a textile yarn but is also used for cigarette filters and some pipe applications and lacquer additives. You'll recall from chapter I that aspirin also is a derivative of acetic acid. It's the *acetyl* group in acetyl salacylic acid (Fig. 1–21).

Properties and Handling. Acetic acid has the strong, pungent smell of vinegar. It's a colorless liquid that is soluble in water and most organic solvents. The concentrations of the commercial grades vary; some are as low as 6%. The USP glacial acetic is 99.5% pure.

Acetic acid melts at 62°F, so shipping poses a special problem. Cool weather can cause the compound to freeze or expand and can rupture a container of pure grades. Tank cars and trucks must be specially lined because of the reactive nature of this acid. Aluminum drums also are used for smaller quantities. The white hazardous (corrosive) shipping label is required.

Acrylic Acid

No commercial market is available for acrylic acids because they never get isolated. These acids are used exclusively to make acrylates, and the processes are completely integrated. The output from acrylic-acid plants is piped directly to the acrylate facilities.*

Once, the only commercial method to produce acrylic acids was the Reppe process based on acetylene. That operation was replaced in the 1960s with stepwise oxidation of propylene to acrolein and then to acrylic acid (Fig. 15–1). Numerous variations ensued, but propylene oxidation accounts for 85% of the acrylic acid.

Acrylic acids eventually end up in the production process for acrylic fibers or latices. The latices are used to make coatings: paints, paper sizing, textile finishes, floor polishes, and adhesives.

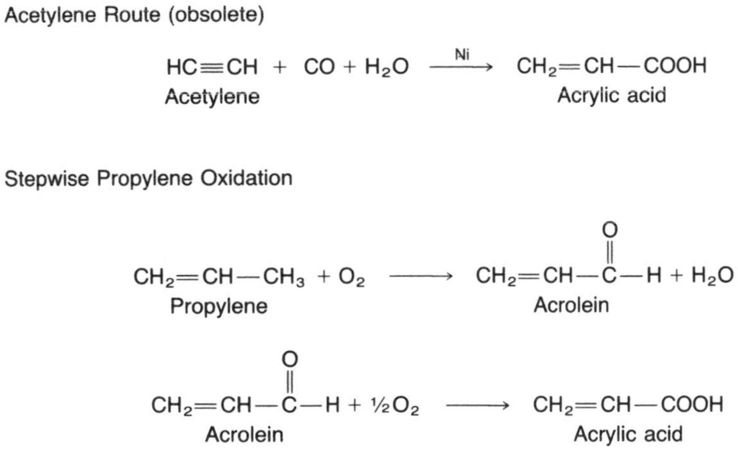

Fig. 15–1 Processes to form acrylic acid

---

*For more information on acrylic acids, see chapter XVI.

Methacrylic acid is only a cousin to acrylic acid—its parents are acetone and hydrogen cyanide. Methacrylic acid produces methacrylates and polymethacrylates. The best known polymethacrylate is Plexiglass.

Adipic Acid

Adipic acid is to Nylon 66 what cumene is to phenol. About 95% of adipic acid becomes Nylon 66, most of which is used to manufacture carpet. Nylon 66 fibers are tough, durable and abrasion-resistant, and they are easy to color, which gives them a secure place in the fiber market (and on the floor).

Adipic acid is produced by oxidizing cyclohexane. The two-step process shown in Fig. 15–2 is used for most production. Cyclohexane is oxidized with air over a cobalt-naphthenate catalyst to give a mixture of cyclohexanol and cyclohexanone. These two products are separated from the unreacted cyclohexane and are hit with a 50% nitric-acid solution over a catalyst. This solution opens the C_6 rings, and adipic acid is formed.

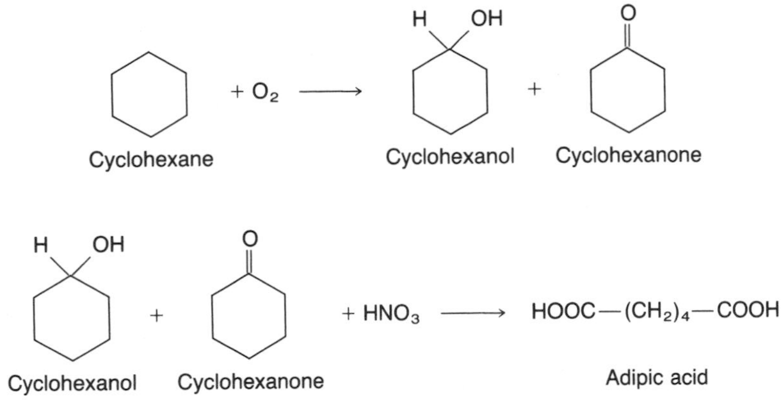

Fig. 15–2   Adipic acid process

A modification of this method is the direct, liquid-phase, air oxidation of cyclohexane in acetic acid with a cobalt-acetate catalyst. The reaction goes directly to adipic acid in 70% yields. However, since the byproduct yield is very high, the process has had limited acceptance.

Properties and Handling. Adipic acid doesn't physically fit the usual image of an acid. Its melting temperature is 306°F. At normal temperatures, this acid is a yellow, crystalline powder that's transported in 1-ton cartons and in drums and 50-pound bags. Adipic acid is only slightly soluble in water, but it dissolves in alcohol. The commercially traded grade is 99.5% pure.

Phthalic Acids

The phthalic acids are made from the three xylenes, ortho, meta, and para. The strange spelling in the name comes from shortening the obsolete form, *naphthalic* acids. Naphtha originally came from the Iranian word for naphtha, which was pronounced *neft*. The three phthalic acids are shown in Fig. 15–3. The major end uses of these *dicarboxylic* acids

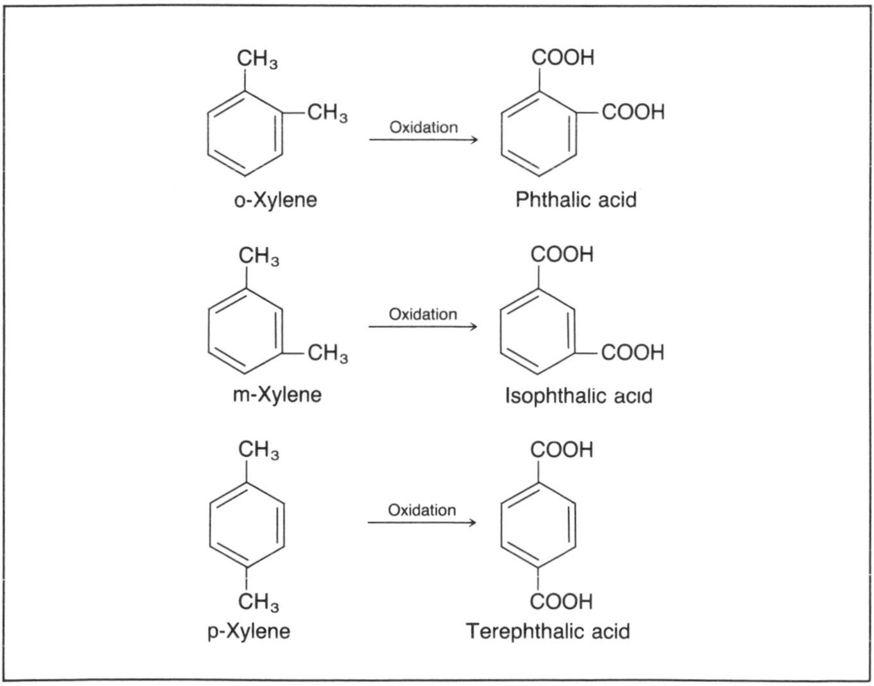

Fig. 15–3 Xylenes and phthalic acids

(two —COOH groups in each) are plasticizers for polymers, alkyd and polyester resins, and fibers.*

Phthalic Acid and Phthalic Anhydride

The commercial product from ortho-xylene is not phthalic acid but the anhydride form, phthalic anhydride. That's because the oxidation of ortho-xylene requires high temperatures. When the phthalic acid forms, it loses a molecule of water, creating the anhydride (Fig. 15–4).

Phthalic anhydride also is made by oxidizing naphthalene, a coal byproduct. The transition of the acid to the anhydride in this process is the same as that in the ortho-xylene method. Naphthalene was the principle precursor for phthalic anhydride until the late 1950s when xylene technology was commercialized. Naphthalene accounts for only 30% of current production.

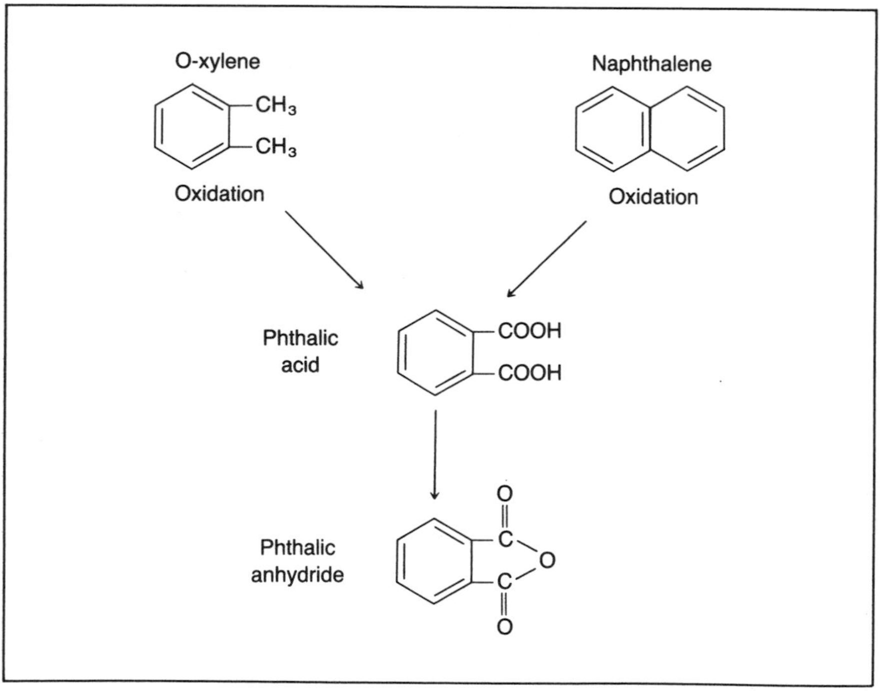

Fig. 15–4   Phthalic anhydride processes

---

*These applications are discussed further in chapters XVIII–XX.

A typical process scheme for phthalic anhydride begins by mixing o-xylene with excess preheated air. This gaseous mixture is then fed to a reactor, consisting of tubes packed with a vanadium-pentoxide catalyst on a silica gel. Like most oxidation reactions, this one is exothermic and heat must be removed from the tubes to maintain the reaction temperature at 700–850°F. Contact time between the reactants and the catalyst is about 0.1 seconds.

The reaction gases (mainly phthalic anhydride, carbon dioxide and water) are cooled, condensed, and purified in stainless-steel facilities. The purified phthalic anhydride (99.5%) is flaked and packaged for shipping. Minor amounts of the byproducts maleic anhydride and benzoic acid also are produced.

Properties and Handling. Phthalic anhydride's melting point is high, 269°F. Therefore, at normal handling temperatures it is a white crystalline solid. It is slightly soluble in water and is commercially available in two grades: pure (99.9%) and technical (99%). Solid phthalic anhydride is shipped in drums and bags; the liquid is transported in heated tank cars and trucks. It is not classified as a hazardous material because it is not corrosive or flammable.

Applications. Phthalic anhydride is used primarily to make plasticizers for polyvinyl chloride. It's also a feed for alkyd resins and polyesters. Minor applications include dyes, fire retardants, esters, and modifiers for drying oil.

Terephthalic Acid

The sole use for para-xylene is to make terephthalic acid (TPA) and its methyl derivative, dimethyl terephthalate (DMT). When DMT is copolymerized with ethylene glycol, it is called polyethylene terephthalate. The commercial name is polyester; a brand name is Dacron.

The acid, TPA, is not unstable like phthalic acid. It cannot dehydrate to the anhydride because the acid groups, —COOH, aren't in the right places. Therefore, TPA is produced, isolated, and traded as an acid.

The original process from p-xylene to TPA was oxidation in the presence of nitric acid. However, handling problems in the plant and nitrogen contaminants in the product led to the demise of that method. Newer technology involves liquid-phase air oxidation of p-xylene. The reaction occurs in an acetic-acid solvent at 400°F and 200 psi, with a cobalt-acetate/manganese-acetate catalyst and a sodium-bromide promoter. The reaction time is about 1 hour. Yields are 90–95%. Solid TPA crystals drop out of the solution as they form. They are removed continuously by filtering a slipstream from the reactor bottom. Crude TPA is purified by aqueous methanol extraction, which gives better than 99% pure flakes.

Properties and Handling. Terephthalic acid has a high melting point, 572°F. At room temperature, it's a white, crystalline solid, insoluble in water or acetic acid. This acid is commercially available in fiber grade (99%) and technical grade (97%). Just to confuse things, the pure, fiber grade of TPA is referred to as PTA. It's routinely shipped in bags, drums, and hopper cars as flakes. No hazardous shipping label is required.

Applications. About 95% of the TPA is used to make polyester. Most of that goes into fiber production; however, some is used in films (magnetic tapes, photographic materials, and electrical insulation). Minor applications for TPA are as herbicides, adhesives, printing inks, coatings, and paints. Polybutylene terephthalate is a new molding resin used as an engineering plastic.

The process to make fibers is either directly through PTA or by conversion first to DMT. That requires reacting PTA with methanol, which gives the ester shown in Fig. 15–5, DMT.

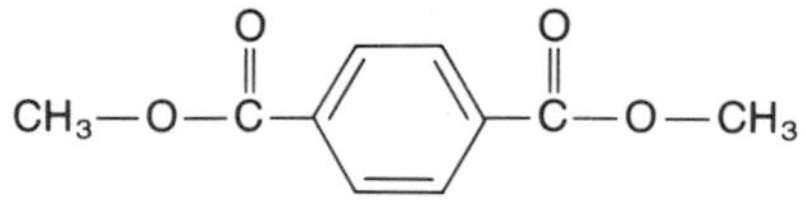

Fig. 15–5 Dimethyl terephthalate

Isophthalic Acid

The stepsister of the other phthalic acids is isophthalic acid, made from meta-xylene. The applications are similar, but the commercial demand is smaller. If it weren't for the fact that m-xylene is a coproduct of the other xylenes, no one would have invented isophthalic acid. The other xylenes and phthalic acids probably would have sufficed.

Isophthalic acid is made by the same process as TPA, liquid-phase air oxidation. Yields are about 80%. Isophthalic does have some redeeming value—it enhances to some extent the mechanical and temperature properties of polyesters, alkyd resins, and glass reinforced plastics.

XVI.

ACRYLONITRILE AND THE ACRYLATES

I am a Bear of Very Little Brain, and long words Bother me.

Winnie-the-Pooh,
A. A. Milne (1882–1956)

There's only a very weak link between acrylonitrile and the acrylates. You might think that acrylonitrile is a precursor of the acrylates. In fact it is, but acrylonitrile is only one of numerous routes to the acrylates and almost an obsolete one at that.

The *acryl* in acrylates and acrylonitrile comes from the Latin *acer* (sharp) and *olere* (to smell). The word acrylic was originally built on the root word *acrolein*, a chemical with a very pungent odor that also is a precursor of acrylates.

The *nitriles* are a group of compounds that can be thought of as derivatives of hydrogen cyanide, HCN. The hydrogen is removed and is replaced by an organic grouping. With *acrylonitrile*, the replacement is the *vinyl* grouping, $CH_2\!=\!CH\!-$, the same one encountered in styrene and vinyl chloride. Acrylonitrile, then, has the formula $CH_2\!=\!CH\!-\!CN$. This compound was originally called vinyl cyanide, but that was before the petrochemical business had good public relations people.

The original route to acrylonitrile was the catalytic reaction of HCN with acetylene. That process combined two compounds that together had all the characteristics you'd like to avoid: poisonous, explosive, corrosive, etc. However, during World War II, acrylonitrile became very important as a comonomer for synthetic rubber (nitrile rubber). Later, the growth of acrylonitrile came from synthetic fibers like Orlon, Acrylon, and Dynel.

Like most acetylene technology in the 1960s, the HCN/C_2H_2 route to acrylonitrile gave way to propylene ammoxidation. That word, *ammoxidation*, looks suspiciously like the contraction of two more familiar terms, ammonia and oxidation, and it is. Standard Oil of Ohio (Sohio) developed a one-step vapor-phase catalytic reaction of propylene with ammonia and air to give acrylonitrile.

$$CH_2\!=\!CH\!-\!CH_3 + NH_3 + \tfrac{3}{2}O_2 \longrightarrow CH_2\!=\!CH\!-\!CN + 3H_2O$$

A byproduct, hydrogen cyanide, is also formed, but there generally is a ready market for it.

Plant Description

The early ammoxidation plants were a two-step design. Propylene was catalytically oxidized to *acrolein* (CH_2=CHCHO). The acrolein was reacted with ammonia and air at high temperatures to give acrylonitrile. Most of this hardware has been replaced by the one-step process.

The Sohio technology is based on a catalyst of bismuth and molybdenum oxides. Subsequent improvements came from using bismuth phosphomolybdate on a silica gel and recently antimony-uranium oxides. Each change in catalyst was motivated by a higher conversion rate per pass to acrylonitrile.

The propylene stream shown in Fig. 16–1 can be either refinery grade (50–70% propylene) or chemical grade (90–95%). Propylene, ammonia, and oxygen are fed in a ratio of 1:1:2 to the vessel containing the catalyst. The vessel is called a *fluidized-bed* reactor because the catalyst moves about like a fluid. The catalyst is usually a very fine, hard powder that flows easily. As the reactants pass through the vessel, they are mixed with the catalyst. Because the catalyst particles are small and abundant, the total surface area exposed to the gaseous or liquid reactants is huge. Therefore, the yields from fluidized-bed reactors are generally higher than from fixed-bed. The main disadvantage is the catalyst loss because mechanically separating the particles out is difficult after the reaction.

Since it is highly exothermic, heat is removed continuously from the reactor by heat exchangers. The residence time of the reactants is about three seconds, then the effluent is quickly cooled and sent to be separated. The unreacted ammonia must be scrubbed out by passing the effluent through a slightly acidic aqueous solution of ammonium sulfate. The gases (unreacted propylene, carbon dioxide, and oxygen) go overhead. The remaining products go with the aqueous solution to be separated in a series of fractionating columns. The major byproducts are acetonitrile and hydrogen cyanide.

The major difficulties with these processes are associated with controlling heat removal from the reactor; stabilizing the catalyst, both mechanically and chemically; and preventing catalyst loss. The latter two problems are caused by using the fluidized bed reactor. Acrylonitrile yields from this process are over 70%, based on propylene feed.

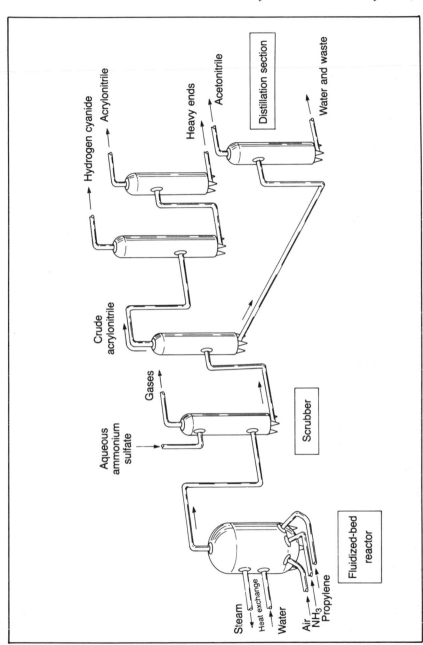

Fig. 16–1 Acrylonitrile plant

Other Processes

New catalysts have been developed, so propane feedstocks can be used instead of propylene. The temperatures are higher (about 950°F) and the residence times are much longer (about 15 seconds), so existing hardware generally cannot handle the new technology. The yields based on propane are also about 70%, but this production has not increased the incentive to build new plants that would shut down existing technology.

Commercial Aspects

Uses. Acrylic fibers account for about half of the acrylonitrile production. Orlon, Acrylon, and Dynel are polymers and copolymers of acrylonitrile. Nitrile rubber has declined in importance but has been replaced by styrene-acrylonitrile (SAN) copolymers and acrylonitrile-butadiene-styrene (ABS) *terpolymers.* These plastics are relatively inexpensive, tough, and durable.

A more recent use of acrylonitrile is to make *adiponitrile* ($NC-CH_2CH_2CH_2CH_2-CN$), a feedstock to Nylon 66 production. Acrylonitrile also is a good treatment for cotton, making it resistant to mildew, heat, and abrasion and more receptive to dyes.

Properties and Handling. Acrylonitrile is a colorless, flammable liquid with a boiling point of 171°F. It is traded commercially as technical grade (99%) and is bulk shipped in lined tank cars or trucks with the hazardous material markings. The linings are necessary due to the corrosive nature of acrylonitrile.

Methacrylonitrile

Methacrylonitrile (MA) can be produced in acrylonitrile plants. This slightly different molecule, $CH_2{=}\overset{\displaystyle CH_3}{\underset{\displaystyle |}{C}}{-}CN$, is copolymerized with acrylic acid, styrene, maleic anhydride, or isoprene to produce a variety of plastics and coatings.

Acrylates and Methacrylates

Acrylates and methacrylates are fun to read about because they end up in all sorts of products you're familiar with, but probably never thought too much about. To begin with, you must understand what *acrylates* are.

An alcohol (a compound with an —OH signature) reacted with an organic acid (one with a —COOH signature) produces an ester (the —COOR signature). This process is called *esterfication*. If the organic acid you use is acrylic acid, the ester is called an acrylate. If the alcohol is, say, methyl alcohol, then the product is *methyl acrylate* but not methacrylate. If you start out with *methacrylic acid*, you get a methacrylate. Finally, if you use methyl alcohol and methacrylic acid, you get *methyl methacrylate*, which is a big star in petrochemicals.

You'll recall from chapter I that the letter R is used as a substitute for a radical, like methyl or ethyl. Therefore, the general equation for esterfication of acrylic acid is as follows:

$$CH_2{=}CH{-}\overset{\overset{\textstyle O}{\|}}{C}{-}OH + ROH \longrightarrow CH_2{=}CH{-}\overset{\overset{\textstyle O}{\|}}{C}{-}OR + H_2O$$

        Acrylic acid   Alcohol   Acrylate

Specifically for the reaction with methyl alcohol, it is written as shown:

$$CH_2{=}CH{-}COOH + CH_3OH \longrightarrow CH_2{=}CH\overset{\overset{\textstyle O}{\|}}{C}{-}OCH_3 + H_2O$$

                                       Methyl acrylate

Now to make sense out of all that, you've got to go to the acid story.

Acrylic Acids

Many petrochemicals have been harnessed because they have two common characteristics: they are simple and reactive. Acrylic acid (AA) is the simplest organic acid *that contains a double bond.* It's that vinyl radical, $CH_2{=}CH{-}$, the same one found in acrylonitrile. Because it is an acid and has the double bond, this compound is very reactive. It readily undergoes polymerization (reacts with itself because of the double bond) and esterfication (reacts with alcohol because it's an acid).

Acrylic-acid use can be traced at least as far back as about 1900. Then it was an additive for paints and lacquers. Since acrylic acid tends to polymerize at low temperatures, it accelerated the drying process of these products. The users probably didn't understand the chemistry of polymerization at the time, only that it worked.

Early routes to acrylic acid were complex and expensive. In 1927, the ethylene chlorohydrin process was introduced. However, this method was still expensive, and little commercial interest was stimulated in the acid. In 1940, a process literally came right off the farm—pyrolysis of lactic acid, a dairy waste found in sour milk. This method improved the economics of acrylic acids because of the availability of a zero-cost raw material, the lactic acid. However, the operating costs remained too high for rapid commercialization. It wasn't until the 1950s, when the Reppe process to acrylic acid was developed, that acrylics production began to take off.

Today, there are numerous commercial routes to the acrylates, making it different than almost any other petrochemical discussed in this book. Several methods bypass the acrylic-acid step and go directly to the acrylate. Five of the most widely used are shown in Table 16–1. The routes involving acetylene and propylene are predominate. Acetylene technology is being replaced by propylene facilities as the industry grows.

Catalytic Oxidation of Propylene

The newest, commercially successful process introduced in the late 1960s changes propylene to acrolein to acrylic acid to various acrylates. Chemical-grade propylene (90–95% pure) is mixed with steam

TABLE 16–1
Commercial Acrylates

Methyl acrylate $\quad CH_2{=}CH{-}\overset{\displaystyle O}{\overset{\|}{C}}{-}O{-}CH_3$

Ethyl acrylate $\quad CH_2{=}CH{-}\overset{\displaystyle O}{\overset{\|}{C}}{-}O{-}CH_2CH_3$

Butyl acrylate $\quad CH_2{=}CH{-}\overset{\displaystyle O}{\overset{\|}{C}}{-}O{-}CH_2CH_2CH_2CH_3$

Isobutyl acrylate $\quad CH_2{=}CH{=}\overset{\displaystyle O}{\overset{\|}{C}}{-}O{-}CH_2\underset{\displaystyle CH_3}{CHCH_3}$

2-ethyl hexyl acrylate $\quad CH_2{=}CH{-}\overset{\displaystyle O}{\overset{\|}{C}}{-}O{-}CH_2\underset{\displaystyle C_2H_5}{CH}(CH_2)_3CH_3$

and air and is oxidized to acrolein (CH_2=CH—CHO). The reaction occurs at 650–700°F and 60–70 psi over a molybdate-tellurium metal-oxide catalyst. (This is standard acrolein technology. Acrolein is used in numerous processes, including the manufacture of allyl alcohol (CH_2=CHCH$_2$OH), the starting material for manufacturing glycerine.)

Acrolein is passed through a second oxidation process to form acrylic acid (see Fig. 16–2). The reaction takes place at 475–575°F, over a tin-antimony oxide catalyst. There are a few byproducts, such as formic acid (HCOOH), acetic acid (CH_3COOH), and carbon monoxide and dioxide. However, the conversion of propylene to acrylic acid is high— about 85%.

Acetylene to Acrylate

The Reppe process, commercialized in the 1950s, involves reacting acetylene, carbon monoxide, and an alcohol (methyl, ethyl, etc.) to give an acrylic ester (an acrylate). The process is carried out at 125°F and 15–30 psi in a nickel-carbonyl/aqueous-hydrochloric-acid solution. The nickel carbonyl acts as both a catalyst and a secondary source of carbon monoxide.

Acrylic acid also can be made from this method by leaving out the alcohol and modifying the operating conditions. The conventional esterfication reaction to produce acrylates can then be run. The lower molecular-weight acrylates (methyl and ethyl) usually are produced using the direct technology. The higher molecular-weight acrylates are made from methyl or ethyl acrylate by what chemists call the *transesterfication reaction*. The higher weight alcohol does a little square dance with the acrylate, changing partners by replacing the methyl or ethyl group with a higher molecular-weight radical.

Reppe process yields are about 80%, but the usual acetylene drawbacks are present: hazardous materials handling and higher costs for raw materials. Nonetheless, as late as the 1970s, the Reppe process accounted for half of the acrylate production.

Acrylonitrile Hydrolysis

The cost to produce acrylonitrile decreased when the ammoxidation process was introduced in the 1960s. Then it became economic to produce methyl and ethyl esters of acrylic acid by hydrolyzing acrylonitrile in the presence of alcohol. Hydrolysis and esterfication occur simultaneously in the presence of sulfuric acid at about 225°F. Yields are approximately 95%. The process consumes the sulfuric acid

Catalytically Oxidizing Propylene

$$CH_2{=}CH{-}CH_3 + O_2 \longrightarrow CH_2{=}CH{-}CHO + H_2O$$
Propylene   Acrolein

$$CH_2{=}CH{-}CHO + \tfrac{1}{2}O_2 \longrightarrow CH_2{=}CH{-}COOH$$
Acrolein   Acrylic acid

$$CH_2{=}CH{-}COOH + CH_3OH \longrightarrow CH_2{=}CH{-}COOCH_3 + H_2O$$
Acrylic acid   Methanol   Methyl acrylate

Acetylene to Acrylate

$$HC{\equiv}CH + CO + C_2H_5OH \longrightarrow CH_2{=}CH{-}COOC_2H_5$$
Acetylene   Ethanol   Ethyl acrylate

Acrylonitrile Hydrolysis

$$CH_2{=}CH{-}CN + 2H_2O + 2H_2SO_4 \longrightarrow CH_2{=}CH{-}COOH + (NH_4)HSO_4$$
Acrylonitrile   Acrylic acid   Ammonium
                                                                      bisulfate

Beta-Propiolactone Process

$$CH_3COOH \xrightarrow{\text{Pyrolysis}} CH_2{=}C{=}O + H_2O$$
Acetic acid   Ketene

Fig. 16–2   Routes to acrylic acids and acrylates

$$CH_2=C=O + CH_2O \longrightarrow CH_2-CH_2-C=O$$

Ketene   Formaldehyde   Beta-propiolactone

$$CH_2-CH_2-C=O + CH_3OH \longrightarrow CH_2=CH-COOCH_3 + H_2O$$

Beta-propiolactone   Methanol   Methyl acrylate

Cyanohydrin Process

$$CH_2-CH_2 + HCN \longrightarrow CH_2-CH_2$$

    Ethylene oxide   Hydrogen   Ethylene cyanohydrin
                cyanide

$$\overset{OH}{C}H_2-\overset{CN}{C}H_2 + H_2O + H_2SO_4 \longrightarrow CH_2=CH-COOH + (NH_4)HSO_4$$

Ethylene cyanohydrin   Acrylic acid   Ammonium bisulfate

Oxidative Carbonylation of Ethylene

$$2CH_2=CH_2 + 2CO + O_2 \longrightarrow 2CH_2=CH-COOH$$

           Ethylene   Acrylic acid

Fig. 16–2 continued

as an ammonium-bisulfate waste. Therefore, this method is expensive, and propylene oxidation remains the preferred route.

Beta-Propiolactone Process

The beta-propiolactone process picks up its classy name from an uncommon intermediate compound. This method really is more involved than the name suggests:

1. Acetone or acetic acid is pyrolyzed to *ketene* (CH_2=C=O);
2. Ketene is reacted with formaldehyde in the presence of an aluminum-chloride catalyst to give beta-propiolactone;
3. Esterfication of beta-propiolactone with an alcohol produces an acrylate.

The yield via this process is about 95%. Energy costs (particularly in the pyrolysis section) are high, and the beta-propiolactone is very toxic, which has limited the commercial popularity of this process to about 10% of production.

Cyanohydrin Process

The granddaddy and original commercial process for manufacturing acrylic acids was the reaction of ethylene oxide and hydrogen cyanide, the poison gas, to produce ethylene cyanohydrin. Then reacting water spiked with sulfuric acid produces acrylic acid. This process is obsolete today primarily because it is expensive and handling the cyanide gas is dangerous.

Oxidative Carbonylation of Ethylene

A route not yet commercialized is the reaction of ethylene, carbon monoxide, and air to give acrylic acid. Ethylene is dissolved in acetic acid. The reaction takes place at 275°F and 1,100 psi in the presence of a palladium-chloride/copper-chloride catalyst. Yields are 80–85%. If the byproduct and corrosion problems can be solved, the process will probably catch on.

Commercial Aspects

Uses. The most commercially important acrylates are ethyl-, butyl-, 2-ethyl hexyl-, and methyl-acrylate, in that order. An important feature of the acrylates is that they readily polymerize if exposed to heat, light, oxygen, or peroxides. More important, they polymerize in water to form a *latex*, which is a dispersion of solid particles in water such as latex paints. A little diversion here might give a better understanding of the value of acrylates.

Emulsion polymerization was developed as part of the synthetic-rubber program during World War II. An *acrylate monomer* (an unpolymerized acrylate) added to water is immiscible—it doesn't mix. If you add an emulsifying agent like soap (yes, soap), the acrylate becomes dispersable (miscible) in water. A water-soluble catalyst induces free radical-type polymerization and the acrylate monomer polymerizes. As the polymer molecules grow to the correct weight and size, they can be stabilized. The resulting mixture is called a latex. Add color pigment and you've got the basics for a latex paint.

Acrylic latices (more than one latex) find many uses in the field of coatings. Every amateur house painter appreciates the handling advantages:

1. When exposed to air and light, latex particles polymerizes further to a hard coating at a moderate speed. (It dries fast.)
2. Before it polymerizes (sets up), latex is dispersable in water. (The brushes and the painter can be cleaned easily.)
3. After it polymerizes, the latex coating is stable and resists oxidation. (It's weather resistant and color-fast.)
4. During the drying process for latex, only water vaporizes. With oil-based paints, naphtha or mineral spirits vaporize during drying. (Latex paints don't pollute.)

Handling. Acrylates are traded as technical grade (99% pure), inhibited or uninhibited. Usually they are sold with trace amounts of hydroquinone as an inhibitor. Methyl and ethyl acrylates are toxic and require a hazardous shipping label. However, the flash points of butyl-, isobutyl-, and 2-ethyl hexyl-acrylates are high, so they are considered safe.

Methacrylates

The methacrylates are cousins to the acrylates, but only one member of this family is commercially important: methyl methacrylate (MMA). The most important feature of MMA is that it polymerizes into a transparent or translucent plastic.

Process Description

The route to methyl methacrylate starts with acetone and hydrogen cyanide and has three steps (see Fig. 16–3):

1. Acetone reacts with hydrogen cyanide in the presence of an aqueous solution of sodium hydroxide at 100–150°F to form acetone cyanohydrin;

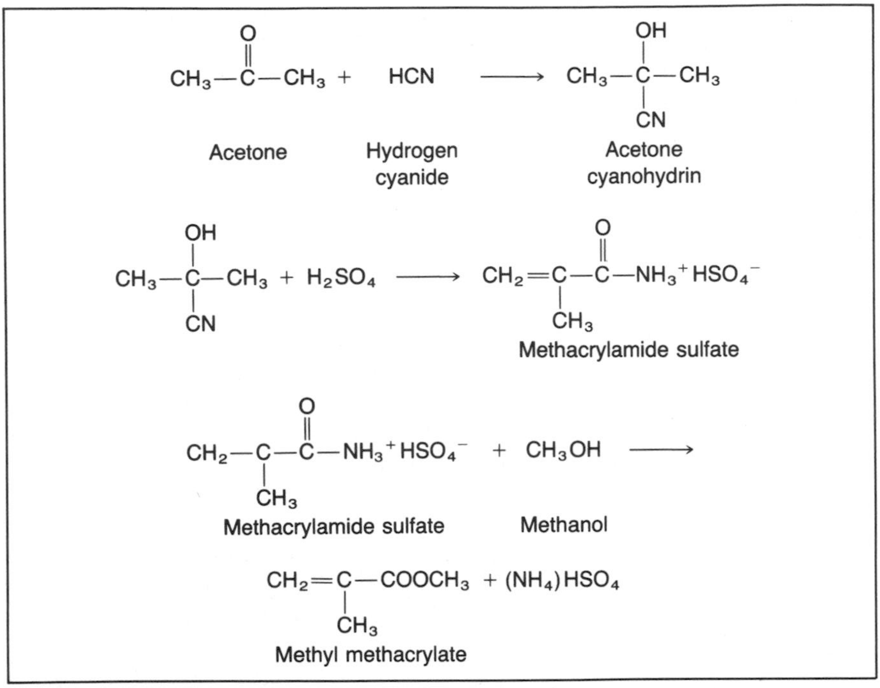

$$CH_3-\overset{\overset{\displaystyle O}{\|}}{C}-CH_3 \ + \quad HCN \quad \longrightarrow \quad CH_3-\overset{\overset{\displaystyle OH}{|}}{\underset{\underset{\displaystyle CN}{|}}{C}}-CH_3$$

Acetone	Hydrogen cyanide	Acetone cyanohydrin

$$CH_3-\overset{\overset{\displaystyle OH}{|}}{\underset{\underset{\displaystyle CN}{|}}{C}}-CH_3 + H_2SO_4 \quad \longrightarrow \quad CH_2=\overset{\overset{\displaystyle O}{\|}}{\underset{\underset{\displaystyle CH_3}{|}}{C}}-C-NH_3{}^+HSO_4{}^-$$

Methacrylamide sulfate

$$CH_2-\overset{\overset{\displaystyle O}{\|}}{\underset{\underset{\displaystyle CH_3}{|}}{C}}-C-NH_3{}^+HSO_4{}^- \ + \ CH_3OH \quad \longrightarrow$$

Methacrylamide sulfate	Methanol

$$CH_2=\overset{}{\underset{\underset{\displaystyle CH_3}{|}}{C}}-COOCH_3 + (NH_4)HSO_4$$

Methyl methacrylate

Fig. 16–3   Methyl methacrylate synthesis

2. Acetone cyanohydrin is reacted with 98% sulfuric acid at 200°F to yield methacrylamide sulfate;
3. The methacrylamide is hydrolyzed and esterfied with methanol at 190–200°F to produce MMA.

After purification, the overall yield is 80–85%.

Commercial Aspects

Uses.   The sole commercial use of MMA is as polymers in various forms: cast sheets, latices, moldings, and extrusions. MMA polymers are best known for their use in the form of clear, transparent sheets with trade names like Plexiglass and Lucite. Applications include advertising signs, aircraft windows, desk tops, lighting fixtures, building panels, and plumbing and bathroom fixtures.

Methyl methacrylate is also used extensively as a copolymer with acrylates in latex paints and as a homopolymer in lacquers. MMA molding and extrusion polymers are used in the automotive industry for control dials, knobs, instrument covers, directional light covers, and tailgate lenses. The last two are probably the largest application of MMA molding powders.

Methyl methacrylate also is used in conjunction with other plastics to achieve translucent or transparent qualities. Transparent bottles, made by copolymerization of MMA with vinyl chloride, are replacing glass containers. Methyl methacrylate has many of the same applications as acrylic latices, and it also can be a comonomer with acrylonitrile to make acrylic fibers.

Properties and Handling. Methyl methacrylate is a colorless, sweet-smelling volatile liquid that boils at 212°F. This compound readily polymerizes with itself and usually has trace amounts of hydroquinone added as an inhibitor. It is traded as technical grade and is shipped in lined tank cars, trucks, and drums. The hazardous material warnings are required on all shipments.

XVII.

MALEIC ANHYDRIDE

How many apples fell on Newton's head before he took the hint?!

Robert Frost (1874–1963)

The unlikely molecule in Fig. 17–1 is a cyclic anhydride known by several names: 2-butene-1,4-dicarboxylic acid anhydride; cis-butene-dioic acid anhydride; maleic anhydride (MA); and, when you've been in the business a long time, "maleic."

The name, *maleic anhydride*, came about in the same fashion as numerous compounds in the petrochemical business. Many organic acids and their derivatives were given common names based on early observations, their special source in nature, or on some special feature of their structure. Maleic anhydride was first isolated in the 1850–75 era by dehydrating *malic* acid, a sugar acid found in apple juice. The Latin word for apple is *malum*, hence, malum, malic, maleic. The suffix, anhydride, which follows each alias of MA, has a simple definition: a compound derived by the loss of a molecule of water from two carboxyl groups (—COOH).

You may wonder why a chemical with such an unusual structure is so popular. The answer, as always, is reactivity. However, in this case the compound is thrice blessed: reactivity is associated with the anhydride grouping (on the right side of Fig. 17–1); with the double bond (on the left); and with the carboxylic acid grouping which (re-) forms when maleic anhydride is mixed with water. That grouping results in an unsaturated dicarboxylic acid that easily undergoes homo- or copolymerization, involving the double bonds or the diacid grouping.

$$
\begin{array}{c}
\quad\quad\quad\quad O \\
\quad\quad\quad\quad \| \\
CH_2 - C \\
\| \quad\quad\quad\quad\quad\diagdown \\
\quad\quad\quad\quad\quad\quad O \\
CH_2 - C \\
\quad\quad\quad\quad \diagup \\
\quad\quad\quad\quad \| \\
\quad\quad\quad\quad O
\end{array}
$$

Fig. 17–1 Maleic anhydride plant

Maleic anhydride can be produced from several feedstocks: benzene, normal butenes or butylenes, and normal butane. The popularity of any one of them has swayed with the economic winds that set the feedstock prices. Benzene was the original choice in the 1940–50 period. Butenes came on strong in the 1950s when this stock became surplus in refineries; but its use faded when the supply fell short again, which stimulated interest in butane. Moreover, normal butane has a yield advantage. If you study the MA molecule, you'll see that it has only four carbons. When benzene (C_6H_6) is the feedstock, two carbon atoms must be eliminated to form the anhydride. These atoms end up as a waste product, carbon dioxide. When butane (C_4H_{10}) is the feedstock, the atoms eliminated are the lightweights, hydrogen. The theoretical yields of maleic anhydride are 1.26 pounds per pound of benzene but 1.75 pounds per pound of normal butane. Therefore, the yield advantage is no small economic factor.

The benzene and butane methods are very similar. Benzene hardware is often adaptable to the butane route because the pressures, temperatures, and even the catalyst are the same. For that reason, benzene plants have been converted to butane plants fairly cheaply.

Process Description

The key to the reactions in Fig. 17–2 is the incredible ability of the catalyst to rearrange the atoms and their bonds. The catalyst, in all three cases, is V_2O_5 (vanadium pentoxide) and a promoter. With benzene, a molybdenum trioxide promoter is added. For butane and butylene, the promoter is phosphorous pentoxide. Consider the changes that occur with benzene in only a single pass through the oxidation reactor, a lapse time of one second:

1. The benzene ring is opened, and two of the carbon atoms are cleaved off as carbon dioxide;
2. The resulting butene molecule undergoes selective oxidation, and the two terminal methyl groups are converted to carboxylic acid groups;
3. A water molecule is lost, giving rise to the heterocyclic anhydride grouping and thus, maleic anhydride;
4. All of this occurs without oxidizing the reactive double bond.

If you set out to accomplish all that in a chemical process, you surely wouldn't expect to be lucky enough to find something as selective and powerful as V_2O_5. Clarke's Third Law is true: "Any sufficiently advanced technology is nearly indistinguishable from magic."

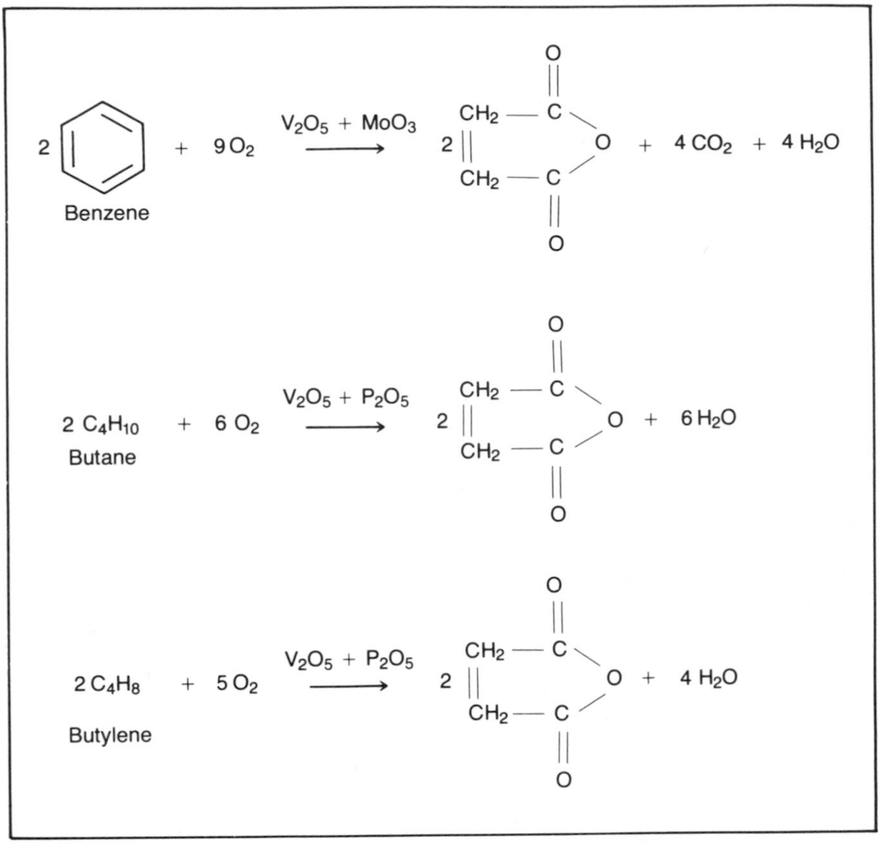

$$2\ \text{Benzene} + 9\,O_2 \xrightarrow{V_2O_5 + MoO_3} 2\ \text{(maleic anhydride)} + 4\,CO_2 + 4\,H_2O$$

$$2\,C_4H_{10}\ \text{Butane} + 6\,O_2 \xrightarrow{V_2O_5 + P_2O_5} 2\ \text{(maleic anhydride)} + 6\,H_2O$$

$$2\,C_4H_8\ \text{Butylene} + 5\,O_2 \xrightarrow{V_2O_5 + P_2O_5} 2\ \text{(maleic anhydride)} + 4\,H_2O$$

Fig. 17–2   Roots to maleic anhydride

The process flow for all three feeds looks like the plant in Fig. 17–3. The feed and compressed air are mixed and vaporized in a heater and are charged to the reactor. The ratio of air to hydrocarbon is generally about 75:1. The reactor consists of a bundle of tubes packed with the catalyst. The reaction temperature is 800–900°F, depending on the feed. The reaction time is extremely quick, so the feed is in contact with the catalyst for only 0.1–1.0 second.

Like most oxidation reactions, this one is exothermic—and extremely so. That's why the catalyst is placed in tubes—coolant is pumped past the tubes to keep the reaction temperature from running away. Also, the reaction is self-sustaining. Once the reaction is hot enough to start, it'll keep going without adding heat.

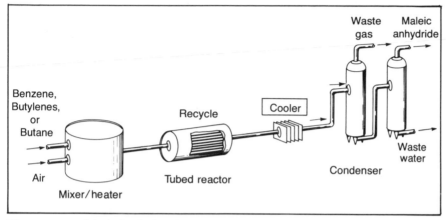

Fig. 17–3   Maleic anhydride plant

The effluent gas from the reactor is passed through a vapor cooler and then into a condenser maintained at temperatures slightly higher than the melting point of maleic anhydride (140°F). The noncondensables (mostly carbon dioxide, benzene, and miscellaneous waste products) are passed to an absorber. There, solvent extraction is used to extract the gases and unreacted benzene is recovered for recycling.

The bulk of the condensables is maleic anhydride, with some maleic acid and water. The acid is readily converted to maleic anhydride by dehydration (removal of a water molecule). This step takes place as the mixture is heated in the final distillation step.

Commercial Aspects

Uses.   About 60% of the maleic anhydride produced is used to make unsaturated polyester and alkyd resins. Polyester resins are used to fabricate glass-fiber reinforced parts. Applications include boat hulls, automobile body parts, patio furniture, shower stalls, and pipe. Alkyd resins are primarily found in coatings (paint, varnish, lacquer, and enamel). Maleic anhydride also is widely used as a chemical intermediate to manufacture plasticizers and dibasic acids (fumaric, maleic, and succinic). Several agricultural chemicals are based on maleic anhydride, the best known being *Malathion*.

Properties, Grades, and Handling.   Maleic anhydride melts at 127°F, so at normal temperatures it is a white solid with an acrid odor. The vapors are highly toxic and will burn your eyes and give you a skin rash. It's soluble in water and in many organic solvents.

Maleic anhydride is available commercially in 99% purity in both molten (liquid, above 127°F) and solid (flakes, pellets, rod, or briquets) forms. It is often shipped in fiber drums or bags. Heated tank cars or trucks are used for liquid shipments. Because of the toxic fumes, the hazardous materials designation must be posted on all shipments.

It is the Beginning of the end.

Charles Maurice de Talleyrand
(1754–1838)
Commenting on the Battle of
Bordino in 1812

Synthesis gas is the underground petrochemical building block. You can't get much more basic than hydrogen and carbon monoxide. But hardly anyone knows as much about it as you do now. The process calls for reacting hydrocarbon (usually methane) with water (using a catalyst) or with oxygen under high pressures and temperatures (without a catalyst).

Some alcohols are made from synthesis gas, including methyl, isobutyl, and normal butyl alcohol. The other alcohols are usually made from olefins. Direct, catalyzed hydration of ethylene yields ethyl alcohol. Isopropyl alcohol is an indirect hydration of propylene operated at lower temperatures and pressures. This method is cheaper than direct hydration. The process to produce 2-ethyl hexanol starts with propylene and synthesis gas, but the dimerization and hydrogenation steps form the alcohol signature group.

The processes for the ketones, acids, acrylonitrile, the acrylates, and maleic anhydride defy simple summarization. You have to read about them individually. The chemical structures for most of them are shown in Table 17–1. It might help you if you try to recognize the signature groups in each molecule.

TABLE 17–1
Chemical Structures of Various Compounds

Name	Nickname	Chemical configuration
Synthesis gas	Syngas	H—H and CO
Methyl alcohol	Wood alcohol	CH_3—OH
Ethyl alcohol	EA	CH_3—CH_2—OH
Isopropyl alcohol	IPA	CH_3—CH—CH_3 with OH below CH
Normal butyl alcohol	NBA	CH_3—CH_2—CH_2—CH_2—OH
2-ethyl hexanol	2-EH	CH_3—CH_2—CH_2—CH_2—CH—CH_2—OH with CH_2—CH_3 below CH
Acetone	DMK	CH_3—C—CH_3 with =O below C
Methyl ethyl ketone	MEK	CH_3—C—CH_2—CH_3 with =O below C
Methyl isobutyl ketone	MIBK	CH_3—C—CH_2—CH—CH_3 with =O below first C and CH_3 below CH
Acrylonitrile	Acrylo	CH_2=CH—CN
Acetic acid	—	CH_3—C—OH with =O below C
Acrylic acid	—	CH_2=CH—C—OH with =O below C
Adipic acid	—	HO—C—$(CH_2)_4$—C—OH with =O below each C

TABLE 17–1 Continued
Chemical Structures of Various Compounds

Name	Nickname	Chemical configuration
Phthalic anhydride	—	
Terephthalic acid	TPA, PTA	HO—C⟨ ⟩C—OH, with O double bonds
Methyl acrylate	—	$CH_2{=}CH{-}C(O){-}O{-}CH_3$
Methyl methacrylate	MMA	$CH_2{=}C(CH_3){-}C(O){-}O{-}CH_3$
Maleic anhydride	Maleic	

The best is yet to come. The most interesting chapters in the book are coming up—the polymers. One reason is that the polymers are a part of your everyday life. Much of what you touch and see nowadays is made in whole or in part from polymers. So you'll keep coming across words that you were familiar with before you ever got into the petrochemical business.

The other reason is that polymer chemistry pulls together a lot of what you've already learned about petrochemicals. So many of them end up in the polymerization process.

The polymers chapters tend to be long. There's a lot to cover under each topic. As a matter of fact, before you get to read about *the* polymers in chapters XIX and XX, you have to read *about* polymers in XVIII. It's a big body of chemistry, but these three chapters should give you a good handle on it.

XVIII.

THE NATURE OF POLYMERS

These are ties which, though light as air, are links of iron.

On Conciliation With America,
Edmund Burke (1729–1797)

Polymers are a pretty complicated subject. That's why they're treated in three successive chapters. In this one you will find a number of ways people classify polymers. It's quite an inventory:

Resins or plastics
Thermoplastics or thermosets
Homopolymers or copolymers
Bifunctional or polyfunctional
Linear or branched or cross-linked
Addition or Condensation

The problem is that polymer chemistry became a virtual explosion of ideas and options as it developed in the 1950s and was further commercialized in the 1960s. There's no easy way to cover polymers other than to wade through. But go ahead. It's not hard and you'll learn a lot.

Polymer History

The first partially synthetic polymer dates back to 1869, when cellulose (wood pulp) was nitrated (nitrocellulose). The cellulose became processible, and by adding camphor (which acted as a plasticizer), it became a clear, tough, moldable product with the trade name "Celluloid." This product was widely used at the end of the 19th century in the form of combs, brushes, photographic film, and shirt collars.

Little commercial development occurred until polymerization chemistry was studied in the 1930s and 40s. The major advancements in polymer commercialization are listed as follows:

1869 Nitrocellulose
1908 Bakelite (first synthetic commercial plastic)
1919 Polyvinyl acetate
1931 Polyacrylates

1936 Polyvinyl chloride
1938 Nylon and polystyrene
1942 Polyethylene and polyesters
1947 Epoxies
1953 Polyurethanes
1957 Polypropylene
1964 Polyimides
1973 Polybutylenes
1977 Linear low-density polyethylene

More than 50% of the chemical industry in the U.S. is now based on or dependent on polymers.

Classifying Polymers

Ask six people in the business to define *resins* and *plastics* and you'll get at least six different answers. Almost everyone will tell you that both are polymers, and that's right. Some may tell you they're interchangeable, but strictly speaking, they're wrong.

Some will tell you that plastics flow when heated or reheated; resins can be heated only once, whereupon they cure or set permanently. A lot more people, particularly in the fabrication end of the business, think resins are unfabricated polymers, and plastics are resins that have been molded or set by a fabrication process.

If you trace the word resin back far enough, you'll find that it was originally defined as a *low* molecular-weight, natural polymer that is an exudate of (it exudes from) vegetable or nonvegetable matter. Examples are rosin (from pine trees), shellac (from insects), and both frankincense and myrrh (aromatic gums from East African and Asian tree species). These resins do not flow like plastics do if heat and pressure are applied. They decompose or melt. (This definition of resin is obsolete in commerce today.)

So you'll get no neat definition of resins and plastics here. But you will know to be careful when someone else uses either term. As for *polymer*, that's defined as a high molecule-weight molecule formed by joining, in a repetitive pattern, one or more types of smaller molecules.

Polymers fall into one of two major classes: *thermoplastics* and *thermosets*. Although thermosets have been around much longer, thermoplastics make up about 80% of industry output. Thermoplastics are linear polymers that can be resoftened several times, usually by applying heat and pressure. They also can be dissolved in solvents suitable for that purpose. That's not true for thermosets. When they are formed (by heat or pressure), these cross-linked, three-dimensional polymers become nonmelting and insoluble. Thermosets actually decompose under heat before they melt.

Both thermoplastics and thermosets can be used in four of the five major application areas: plastics, elastomers, coatings, and adhesives. But only thermoplastics can be used to make fibers. During the spinning and drawing process of fiber processing, it's necessary to orient the molecules. Only unbranched, linear polymers (not thermosets) are capable of orientation.

Polymers result from polymerization—the chemical combination of a large number of molecules of a certain type, called monomers. Monomers can be *bifunctional* (capable of joining with two other monomers) or tri- or *polyfunctional* (joining with three or more monomers). When bifunctional monomers react with each other, linear thermoplastic polymers form. If tri- or polyfunctional monomers enter into the reaction, you get cross-linked polymers, most of which are thermosets. Fig. 18–1 illustrates these variations.

In some cases, the monomers react with themselves to form *homopolymers*:

Ethylene to polyethylene

Vinyl chloride to polyvinyl chloride

Styrene to polystyrene

However, most of the time, two or more different monomers react to form *copolymers*:

Butadiene and styrene to Buna S rubber

Styrene and acrylonitrile to SAN

Ethylene glycol and maleic anhydride to unsaturated polyester

Some copolymer variations also are illustrated in Fig. 18–1.

Making Polymers

The polymerization process can be an *addition* or a *condensation* reaction. Addition involves monomers containing a carbon-carbon double bond ($R—CH=CH—R_1$, where R and R_1 can be hydrogen, chlorine, an alkyl, or an aromatic group). Condensation polymers generally result from simple reactions involving two monomers, each containing different functional groups. A common example is adipic acid and ethylene glycol to make polyester. The two monomers react in such a way that a small molecule like water or methanol is a byproduct.

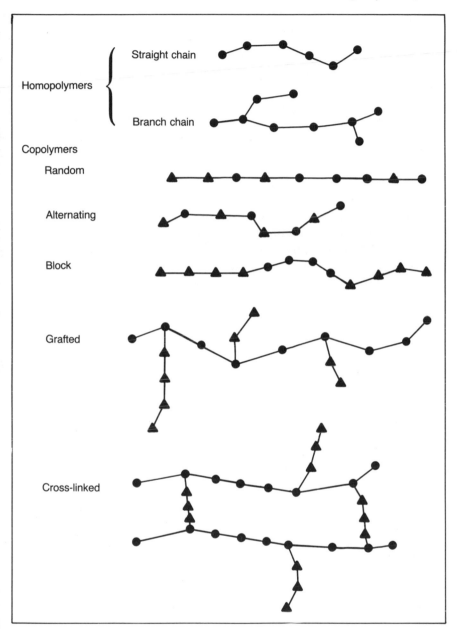

Fig. 18–1 Polymer structures

Addition Polymerization

Addition polymerization is a technique for adding monomers end to end. It involves three steps: initiation, propagation, and termination.

Initiation. The trick to getting the reaction started is to use a catalyst. A typical catalyst is an organic peroxide because these molecules are somewhat unstable. When they're heated, they decompose and, like the benzoyl peroxide shown in Fig. 18–2, turn into one or more free radicals. In this case, they are the benzoyl and the phenyl radicals. As you'll recall, a radical is a nearly complete molecule, but all of the valence requirements are not satisfied. So it is very anxious to meet another molecule. The dot at the end of the radical denotes the unsatisfied valence site, a very reactive location. The phenyl radical in the presence of an abundance of monomers reacts with a monomer molecule. However, since there are not enough atoms (usually hydrogen) to go around, the unsatisfied valence condition transfers to the monomer. This condition is shown in the third equation in Fig. 18–2 and is called **A·** for short.

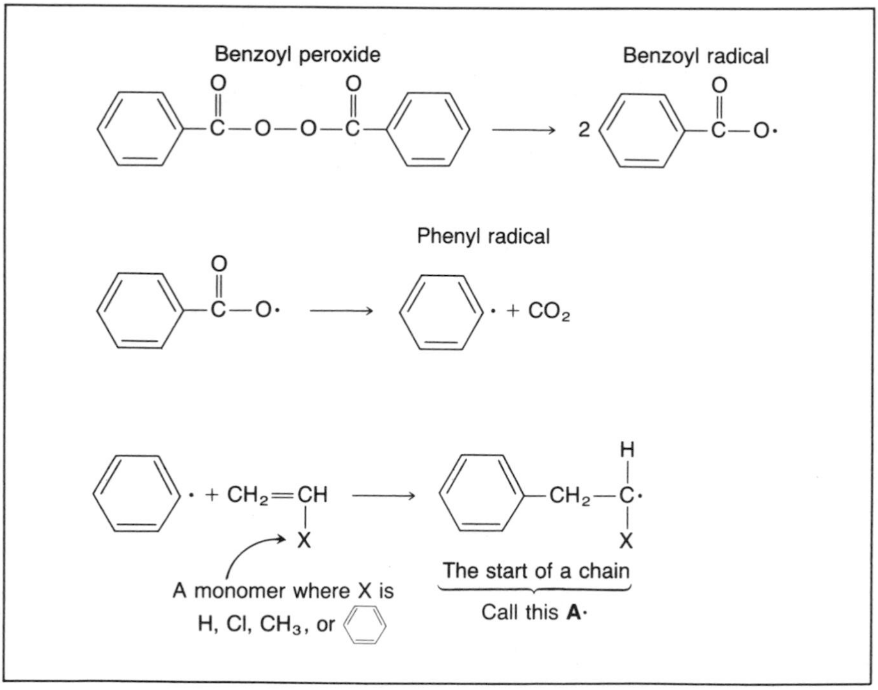

Fig. 18–2 Initiation step

Propagation. The new radical **A·** reacts with another monomer molecule to give a new larger radical, as shown in Fig. 18–3. This chain growth continues until propagation is terminated. The propagation step in a commercial process usually takes a few seconds, and typically a thousand or more monomers form the chain.*

Termination. Several mechanisms can stop propagation. A common way occurs when the monomer concentration is so low that the free radical chains *dimerize.* That is, they collide with each other and form a stable polymer, as shown in the third reaction in Fig. 18–3.

Branch polymers (or *branch chains*) are short or long chains at right angles to the backbone of the original chain. Short chains can be deliberately added using comonomers such as butene-1 or hexene-1. Long-chain branching often happens in high-pressure polyethylene processes. In the propagation step, a growing polymer radical extracts

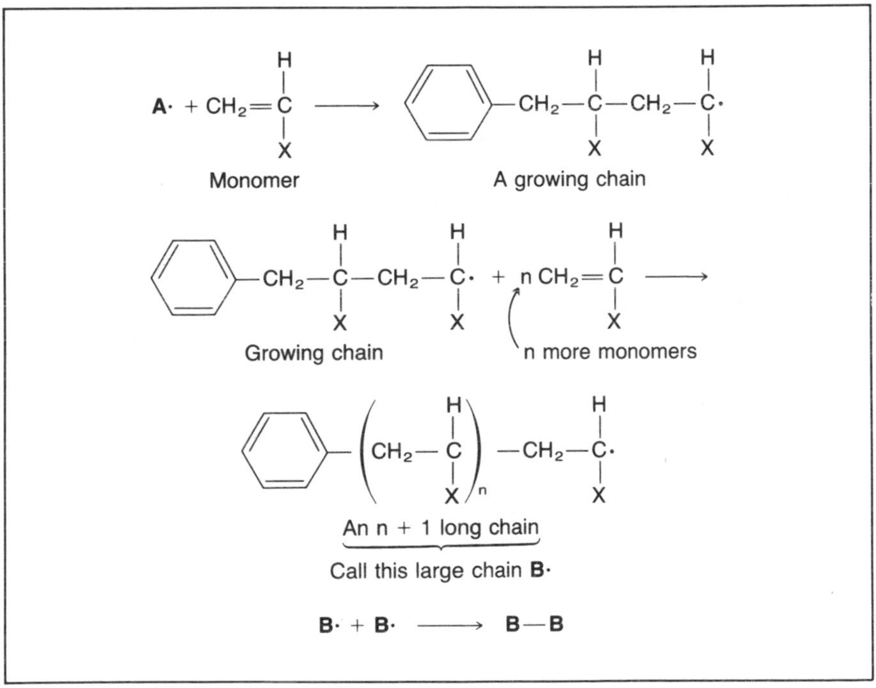

Fig. 18–3 Propagation and termination

---

*Using special catalysts, some polyethylenes are produced with up to 150,000 repeating units.

an inside hydrogen atom from a finished polymer chain (Fig. 18–4). The molecule then becomes a new polymer radical at the extraction site, and a chain can grow there. Sometimes a chain transfer agent is added to facilitate the process. Isobutane, propylene, and dodecyl mercaptan do well. The chain transfer agent gives up a hydrogen to a growing polymer and becomes a radical. Then the chain transfer agent/radical is reactive and extracts the inside hydrogen even more easily than a growing polymer can.

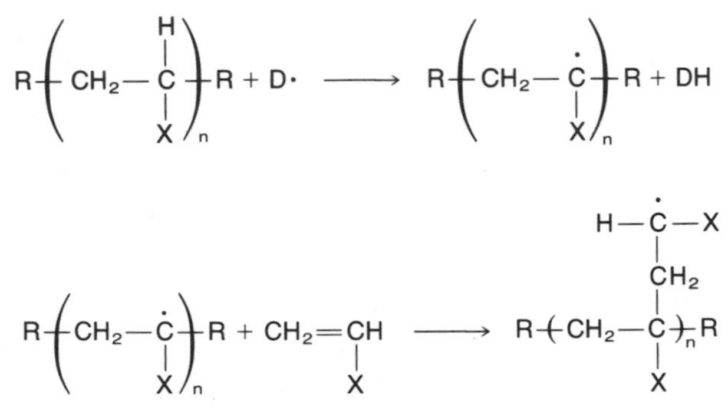

Fig. 18–4   Polymer branching

Cross-linked polymers occur when polymer chains are linked together at one or more points (other than at their ends). Cross-linking can occur when the monomers involved are *polyfunctional.* That is, they have more than two active sities where links can be attached. Therefore, these monomers grow like long chains, but the chains also link with each other. Cross-linking also can be initiated by adding special agents (like Charles Goodyear did when he accidently spilled some sulfur into a vessel of molten natural rubber. In the process, he discovered vulcanization, cross-linking with sulfur atoms.) Cross-linked polymers lose their moldability even when they're reheated because the molecules are chemically bonded in place and do not slip and slide.

The length, branching, and cross-linking of polymers are controlled by timing the three steps. An abundance of catalyst results in an abundance of free radicals. When that happens, the monomer concentration decreases rapidly as a relatively high number of polymers start growing at once. This growth results in early termination and creates many small (low molecular-weight) polymers. The properties of these polymers differ (maybe for better or for worse) from a relatively small number of large (high molecular-weight) polymers. Usually the large molecules are what you're trying to produce.

Copolymers. Mixtures of two or more different monomers can undergo addition polymerization to form copolymers. Why copolymer? Well, polymers have different properties that depend on their composition, molecular weight, branching, and crystallinity. Many copolymers have been developed to combine the best features of each monomer. For example, polystyrene is inexpensive and clear, but it also is brittle without toughness. It needs *internal plasticization*. By copolymerizing styrene with small amounts of acrylonitrile or butadiene, the impact and toughness properties are dramatically improved.

Another reason for copolymerization is to insert functional grouping in the polymer. In that way the polymer can be reacted later with itself or with another monomer to give a cross-linked thermoset.

A third interest in copolymers is crystallinity. Transparency and translucency are greatly affected by crystalline properties, which can be regulated by copolymerization.

Condensation Polymerization

Condensation polymers are always copolymers. They are always formed by a series of chemical reactions involving two reactive sites that join to form a bond. Byproducts are usually given off, generally small molecules such as water, methanol, or hydrogen chloride.

Since two reactive sites are necessary, *bifunctional* monomers are often used in condensation polymerization. A bifunctional monomer includes molecules with two identical signature groups. Examples are terephthalic acid or ethylene glycol, both shown in Fig. 18–5. When one of the bifunctional monomers is used, the polymerization step is end-to-end, forming long chains. The reaction in Fig. 18–5 is a simple esterfication of ethylene glycol with terephthalic acid to make polyethylene terephthalate (polyester fiber).

Numerous bifunctional monomers are used in condensation polymerization. Some more popular signature groups are shown in Fig. 18–6. Important copolymers made by condensation include *epoxies* (from epichlorohydrin and Bisphenol A), nylon (from diamine and adipic acid), polyesters (glycol and anhydride or dicarboxylic acid), polycarbonate (phosgene and glycol or Bisphenol A), and polyimides (diamine and dianhydride). As always there are exceptions, and one is Nylon 6 made by a ring-opening reaction of caprolactam.

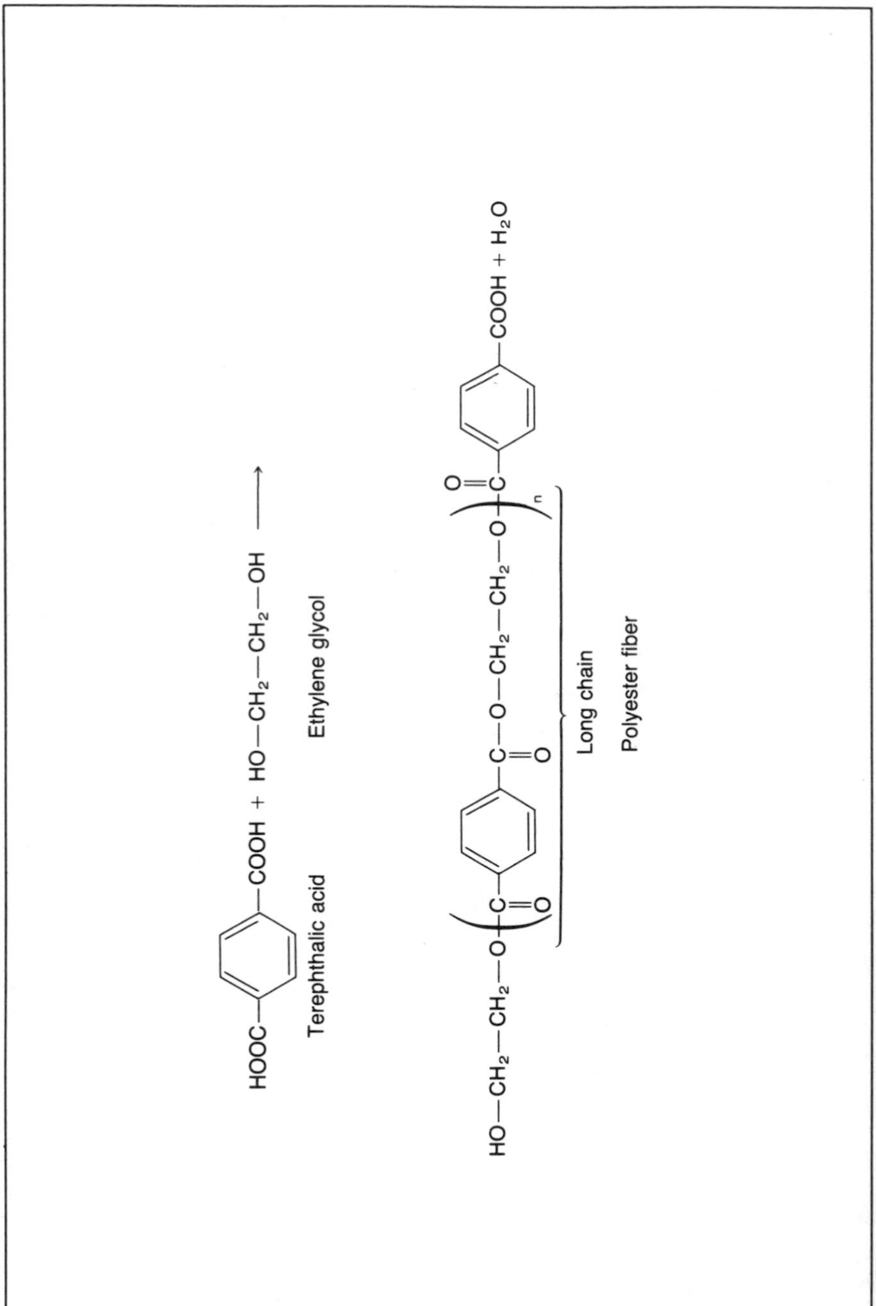

Fig. 18-5. A condensation polymerization

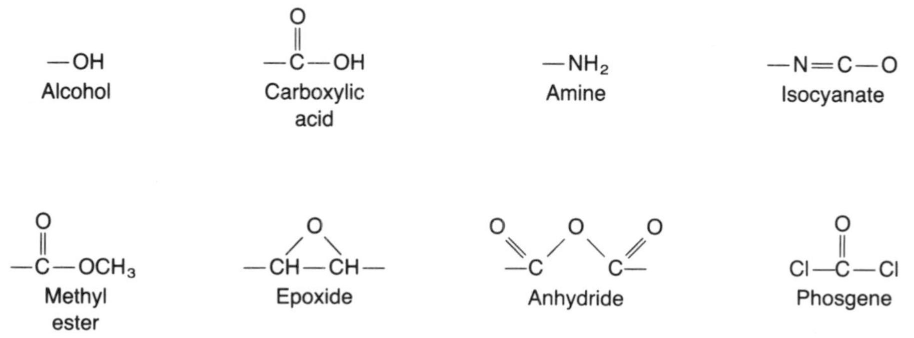

Fig. 18–6 Common signature groups used in condensation polymers

Thermosets

Thermosets cure into nonmelting, insoluble polymers. Frequently, the curing needs heat, pressure, or a catalyst to proceed. Often the final phase, which completes the cross-linking, occurs in the fabrication or molding presses. The chemistry is similar to that for thermoplastics, but there are more reactive sites per monomer. (These monomers are polyfunctional.) Consequently, more three-dimensional cross-linking takes place.

The simplest way to achieve three-dimensional cross-linking is to use monomers with three or more reactive sites. Examples are maleic anhydride, butadiene, isoprene, epichlorohydrin, pyromellitic dianhydride, trimethylol propane, and phenol-formaldehyde prepolymer.

As an illustration, consider some of the *elastomers*. In its natural state, rubber lacks toughness. In the 1939 accident mentioned, Goodyear found that by reacting latex rubber (natural or synthetic) with sulfur, he could improve its strength and toughness and increase its temperature properties. What he was doing was cross-linking rubber with sulfur in a process now commonly called vulcanization. The reaction with polyisoprene rubber is shown in Fig. 18–7. Other synthetic rubbers such

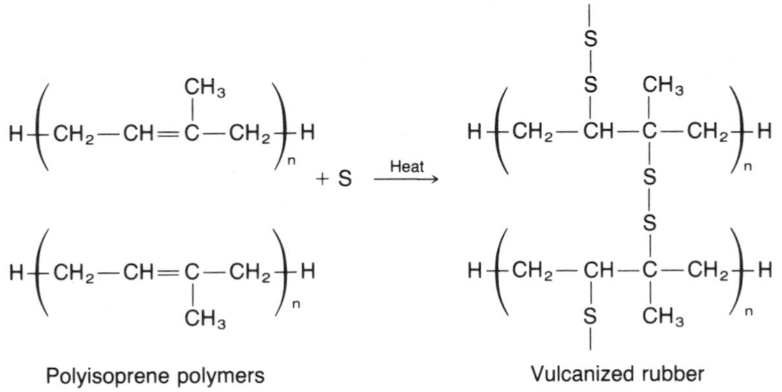

Fig. 18–7   Vulcanization of isoprene rubber

as butyl rubber (from isobutylene and butadiene), Buna S (butadiene and styrene), Buna N (butadiene and acrylonitrile), and neoprene (chloroprene) can be vulcanized to thermoset rubbers. They all have the polyfunctional configuration that makes cross-linking with sulfur possible. After vulcanization, these rubbers are tough, resist deforming, and are insensitive to extreme temperatures. They also resist solvents and are nonconductive. Other important thermosets include phenolics like Bakelite, epoxy resins, polyimides, and polyurethanes.

Polymerization Methods

Most polymers are made by one of four processes commercially available. Each method has advantages related to the monomer and the end use of the polymer.

Bulk Polymerization. Bulk polymerization is the simplest method. Monomers and an initiator are mixed in the reactor shown in Fig. 18–8 and are heated to the desired temperature. The bulk process is more suitable for condensation polymers because the heat of reaction is low. Methyl methacrylate is changed to Plexiglass in the bulk process. High-pressure ethylene polymerization also is done this way. However, other addition reactions frequently become too exothermic—without an adequate heat removal system, the reaction conditions tend to exceed optimum quickly.

Solution Polymerization. Highly exothermic reactions can be handled by solution polymerization. This process uses excess solvent, which absorbs and disperses the heat of reaction. The solvent also prevents slush or sludge formation, which may happen in the bulk process when the polymer volume overtakes the monomer. The solution process is particularly useful when the polymer is to be used in the solvent, say as a coating. This low-pressure process also can produce high-density polyethylene and polypropylene. However, it makes removing residual traces of solvent or catalyst difficult.

Suspension Polymerization. In the suspension process, monomers and an initiator are suspended as droplets in water or in a similar medium. The droplets are maintained in suspension by agitation (active mixing). Sometimes a water-soluble polymer like methylcellulose or finely divided clay is added to stabilize the mixture. After formation, the polymer is separated and dried. This route is used commercially for vinyl-type polymers such as polyvinyl chloride and polystyrene.

Emulsion Polymerization. Soap is usually the agent for emulsion polymerization. The two ends of soap molecules are *oleophilic* (attracted to oil) and *hydrophilic* (attracted to water). The soap molecules form

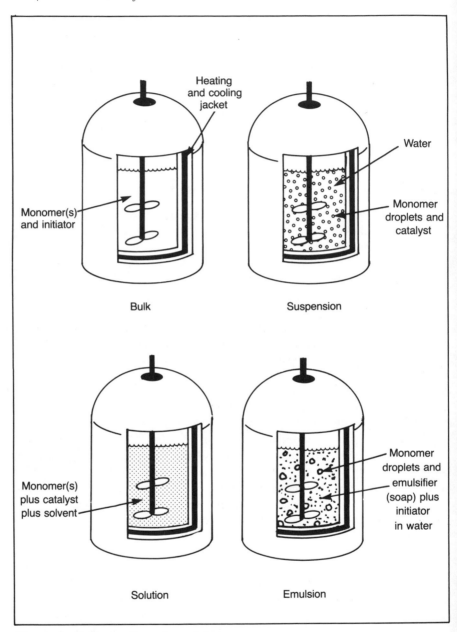

Fig. 18–8 Processes for polymerizing

micelles, tiny structural units with the oleophilic ends pointing inward, holding onto some monomers. The hydrophilic ends point outward, interacting with the water medium. The miscelle remains suspended in the water, and polymerization occurs within the micelle. The mono-mer is suspended in droplets as in the suspension process by active agitation. The oleophilic end of the soap grabs some of the mono-mer and transfers it into the micelle as the polymer grows. Very high molecular-weight polymers are produced by this technique in the form of latex. This process is particularly suitable for polymers used in paints, such as polyvinyl acetate.

Polymer Properties

The proof of the polymer is in its properties. Engineers use physical properties to select polymers. These characteristics include density, tensile and impact strengths, toughness, melt index, creep (ability to elongate), modulus of elasticity, electrical functions, thermal conductivity, appearance, flammability, and chemical resistance. Price and fabricat-ing costs also are considered to select the correct polymer for each application.

Generally speaking, the physical properties of polymers depend on crystallinity, molecular weight, molecular-weight distribution, linearity/cross-linking, and chemical composition or structure.

Crystallinity is one of the key factors influencing polymer char-acteristics. You can think of crystallinity in terms of how well a polymer fits in an imaginary pipe, as in Fig. 18–9. Linear, straight chains do this very well. Bulky groups and coiled and branched chains cannot line up to fit in the pipe. Therefore, they are *amorphous*, the opposite of crystal-line. The spectrum from amorphous to crystalline includes methyl methacrylate, polypropylene, low-density polyethylene, linear low-density polyethylene, high-density polyethylene, and nylon.

With increasing crystallinity, polymers tend to be more dense. They're not too different from pasta. A pound of uncooked spaghetti fits in a smaller box than a pound of macaroni. Along with increasing density comes greater tensile strength, higher softening point, and more opaqueness. Elongation (stretch) and impact strength decrease. Sur-prisingly a more crystalline polymer is less translucent. It doesn't help to ask you if that's crystal clear, because that expression implies just the opposite.

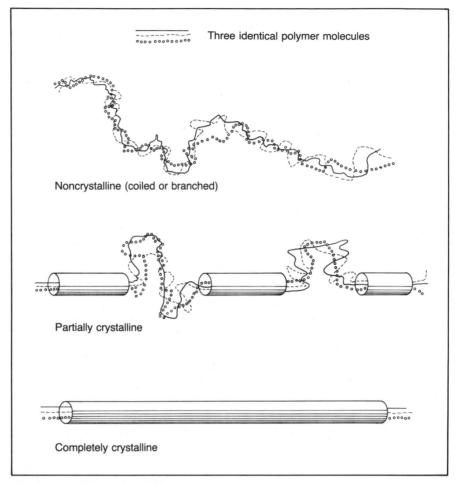

Three identical polymer molecules

Noncrystalline (coiled or branched)

Partially crystalline

Completely crystalline

Fig. 18–9   Polymer crystallinity

Also, molecular weight influences melt viscosity, tensile strength, low-temperature brittleness, and resistance to tearing. The structure and chemical composition of a polymer affect various properties, such as thermal stability, stress-strain, density, flammability, and moisture absorption.

Thermal Stability.   The presence of side chains, cross-linking, and benzene rings in the backbone of a polymer increases melting temperatures. For example, a spectrum of polymers with increasing melting temperatures are polyethylene, polypropylene, polystyrene, nylon, and polyimide.

Stress-Strain Characteristics. Linear chain polymers are very flexible and are subject to creep or stretch. Branching or rings in the backbone have a stiffening effect, for example:

Polyethylene	Soft, tough, high creep
Polypropylene	Hard, tough, medium creep
Polystyrene	Hard, brittle low creep
Cross-linked thermosets	Hard, brittle, no creep

Density. When something other than CH— or CH_2— is introduced to polymers, most of them become more dense. In order of increasing density are polypropylene, polyethylene, polystyrene, polyvinyl chloride, and Teflon.

Flammability. Chlorine, fluorine, bromine, or phosphorous in a polymer reduces flammability. Thermosets are more flame resistant than thermoplastics.

Moisture Absorption. Moisture absorption relates directly to the atoms making up the polymer. More moisture-absorbing molecules have less dimensional stability. Strength, stiffness, electrical properties also are adversely affected.

Those are the generalities of polymers. The specifics of low- and high-density polyethylene, polypropylene, polyvinyl chloride, and polystyrene are covered in the next chapter. Resins and fibers in the last.

XIX.

THERMOPLASTICS

Let there be spaces in your togetherness.

The Prophet, Kahlil Gibran (1883–1931)

Like it or not, it's becoming a plastic world. Plastics are penetrating the traditional markets for paper, cotton, wool, wood, leather, metals, and concrete. The growth of plastics would be even faster if they weren't made out of such an expensive raw material, petroleum. However, many of the materials that plastics replace have important energy components in their creation as well. So the advances in plastics continue.

The big four thermoplastics are polyethylene, polypropylene, polyvinyl chloride, and polystyrene. Like most other thermoplastics, these materials are long chain polymers that become soft when heated and can be molded under pressure. They are linear or branch chained, with little or no cross-linking. Thermoplastic technology is still developing, and research in copolymerization, catalysts, processing, blending, and fabricating continues.

Polyethylene

You have to shake your head in wonder when you think about how the largest selling plastic was developed—by accident. In 1933, the scientists at the ICI labs in England were attempting to make styrene by reacting benzaldehyde with ethylene under high pressure. Instead, their efforts resulted in a reactor lined with a solid, white, wax-like material—polyethylene (Fig. 19–1). Six years later, a German scientist at IC Farben-Industrie, Max Fischer, was trying to synthesize lube oils from ethylene. He used an aluminum powder and titanium catalyst at low pressures and produced the same solid, white wax-like material.

Indicates repeating links

Fig. 19–1   Polyethylene

The English experience eventually developed into the high-pressure polymerization route to low-density polyethylene (LDPE). The German experiment was the forerunner of high-density polyethylene (HDPE).

The next polyolefin to arrive on the commercial scene was polypropylene in 1957 but only after some conscientious, on-purpose research. The most recent process is linear low-density polyethylene (LLDPE), developed in 1977. This method combines some of the best features of both LPDE and HDPE by using butene-1, hexene-1, or octene-1 as a comonomer.

The most important characteristic of polymers is properties. How they look and how they react under various conditions is paramount. That's why different types of polyethylene have been commercialized. Fer example, LDPE is more flexible and has better clarity than HDPE. HDPE has greater strength and less creep, and it is less permeable to gases. Permeability seems to go along logically with density, but it also is associated with with molecular weights, branching, and crystallinity. LLDPE has most of the good features of both LDPE and HDPE: strength, flexibility, clarity, good dielectrics, and high/low temperature stability (for wire and cable shielding).

Despite the names, there's not much difference in the densities of these polyolefins. For each, the density varies according to the degree of polymerization generated in the process. However, in general LDPE is about 0.920–0.935 grams per cubic centimeter (g/cc); HDPE is about 0.955–0.970 g/cc; LLDPE varies between 0.920–0.950. That's a variation of less than ±5%. It's really the molecular weights, chemical structure, and the properties that make the difference, not the densities.

The routes to the polyethylenes also fall into several categories:

1. High-pressure polymerization, using the free radical catalysts, gives LDPE. Reacter pressures run as high as 50,000 psi; temperatures up to 500–650°F.
2. Medium-pressure polymerizations with metal-oxide catalysts, such as chromium or molybdenum oxide produce HDPE.
3. Low- and medium-pressure polymerizations with Ziegler catalysts give HDPE. Ziegler catalysts are trialkly aluminum/titanium tetrachloride, the not-your-everyday-chemicals developed as catalysts by Karl Ziegler in the 1950s.
4. Low-pressure, gas-phase, fluidized-bed polymerization of ethylene with a comonomer and fluid catalyst particles produces LLDPE.

Ironically, the high-pressure process produces a low-density polyolefin; low and medium pressures can form high-density materials. You'd think it would be just the opposite, but it has to do with branching and crystallinity. High pressure combined with the free-radical catalyst promotes branching, which leads to less crystalline molecules —the less crystalline, the less dense. (Recall the pasta example in Fig. 18–9. Uncooked spaghetti is more dense than uncooked macaroni, and the spaghetti-shaped polymers are completely crystalline.)

LDPE Process Description

Since the polymerization process for LDPE requires pressures of 15,000–50,000 psi, the equipment is very expensive. Even worse, the reaction is exothermic and can exceed optimum conditions, which can cause an explosion. In fact, the original developers of LDPE, the ICI laboratories, shut down the development work for several years when their first on-purpose attempt to make polyethylene exploded and destroyed their lab.

A flow diagram of this process is shown in Fig. 19–2. Compressing *polymer-grade* ethylene (99.9% purity) to the reaction pressure is a major step, and several compression stages are needed. Since ethylene begins to polymerize on its own above 212°F, cooling between compression steps is neccessary. (Compression always causes gas temperature to increase.) The pressurized ethylene and catalyst enter a vessel called an *autoclave* reactor.*

In the reactor is a mixture of ethylene and growing polymers. The polymerization reaction is exothermic, so the rate of ethylene added can be regulated to keep the reactor temperature constant. At the same time, polyethylene is constantly drawn off to maintain the balance. This is called running the process under essentially *adiabatic conditions* (no heat needs to be added). However, the water jacket around the autoclave reactor is still needed as a big sponge, sopping up excess heat that variations in the reaction can cause. It's also insurance in case of a runaway.

Residence time of the ethylene in this scheme averages 25–30 seconds. Ethylene conversion is only about 15–20% per pass.

To separate the ethylene in the effluent from the LDPE, the pressure is lowered in successive vessels and the ethylene flashes off (vaporizes). The ethylene is recycled to the compressors. The LDPE, still in a molten (hot-liquid) stage, is cooled and extruded. Then it is pelletized, dried,

---

*An autoclave is any vessel that can be closed and can maintain an increase in pressure. Usually something goes on inside that generates pressure, like a chemical reaction. (A doctor or dentist sterilizes instruments with a steam-generated autoclave.) Also, an autoclave reactor is generally designed to handle both high temperatures and pressures.

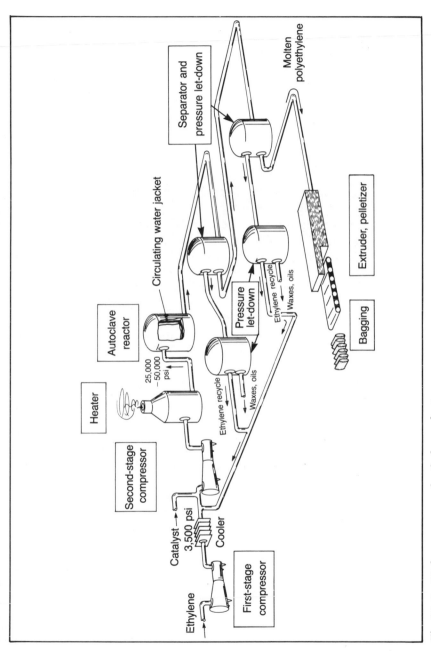

Fig. 19–2 High-pressure polyethylene plant

and bagged. Often mineral oil is used as a carrier for the catalyst. Generally, both are left entrained in the LDPE, and the mineral oil acts as a plasticizer.

HDPE Process Description

In the 1950s, three different processes were developed independently to make HDPE:
1. Solution
2. Slurry or particle form
3. Gas phase

Solution Process. In the solution method, polymerization occurs in a vessel filled with a solvent like cyclohexane (Fig. 19–3). The solvent has several jobs. It keeps the reactants fluid, even following polymerization. It also sponges up much of the heat from the exothermic reaction, and it controls the rate of ethylene consumption.

This reaction also runs in an autoclave reactor. Temperatures of 275–350°F and pressures of 300–450 psi are maintained. Residence times in the reactor commonly average 1.5 hours—it's a slow process.

Only small amounts of the newer catalysts are needed, so catalyst separation from the reactor effluent often is not necessary. Downstream of the reactor are the purification, recycle, and cleanup steps, until the polymer particles (called crumbs) are dried. Then the crumbs are extruded, pelletized, and packaged.

The catalysts used in the reactor (chromic oxide, molybdenum oxide, or trialkyl aluminum/titanium tetrachloride) are very sensitive to contaminants such as water, oxygen, peroxides, or acetylene. For that reason, the ethylene feed must be polymer grade (99.9% purity). The recycle streams, especially the cyclohexane solvent that has been steam washed, must be thoroughly dried. It doesn't take much more than trace amounts of these impurities to poison the catalysts.

Particle-Form Process. The unique-looking equipment in Fig. 19–4 is a loop reactor used in the particle-form method. This process also takes place in a solvent (usually normal hexane), so the mixture can be pumped continuously in a loop during polymerization. Since residence times are longer, this type of reactor produces polyethylene molecules of higher molecular weights that also have high melt temperatures.

Feeds (normal hexane, a comonomer if any, ethylene, and a catalyst) are pumped into the loop and are circulated. Polymerization takes place continuously. A slurry of HDPE in hexane settles in the verticle legs and is drawn off continuously or intermittently.

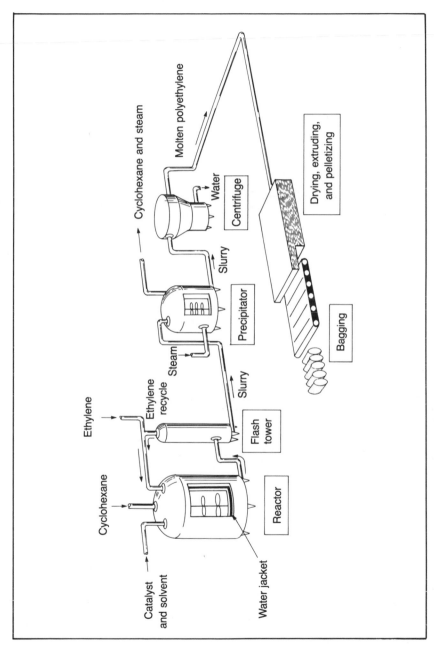

Fig. 19–3   Low-density polyethylene solution process

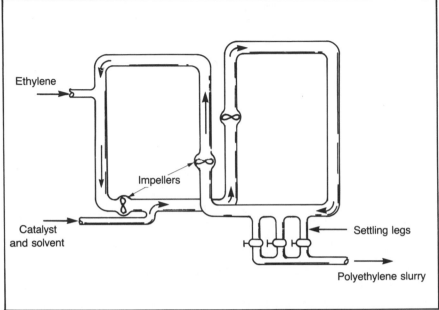

Fig. 19-4 Low-density polyethylene particle-form process

The loops are pipes 10–20 inches in diameter and about 50 feet high, with a total length of 250–300 feet. They hold about 600 cubic feet of slurry and are water jacketed to control the heat. The reaction temperature in the process is less than 212°F, with pressures of only a few hundred pounds. Therefore, this process is more economical (energy saving) than the others previously discussed. When the slurry is withdrawn from the loop reactor, processing continues downstream of the reactor, as shown in Fig. 19–2.

Gas Phase.   No solvent is used in the gas-phase process. Ethylene and a very reactive, silica-supported, chromium-based catalyst are blown into a tall reactor. The ethylene and comonomer (if any) polymerize and drop to the bottom of the vessel, where they are drawn off. The unreacted ethylene goes out the top of the vessel and is recycled. The HDPE is easy to clean, since only the residual ethylene must be separated. Coupling that with low temperatures (185–212°F) and pressure of 325 psi makes the gas-phase method relatively inexpensive.

LLDPE Process Description

You might think of LLDPE as third-generation polyethylene. First there was the high-pressure reaction for making LDPE in the 1930s.

Then in the 1950s, there were the low pressure HDPE processes, made possible by the Ziegler catalysts. In 1977, a method was introduced that permitted low pressures and temperatures, but the polyethylene remained linear and had low density (0.92–0.96 g/cc). The breakthrough used metal catalysts and a comonomer, which allowed the low pressures but still produced low-density polyethylene. About 8–12% of butene-1 or octene-1 was used as the comonomer.

The hardware is much like that the low-pressure gas-phase method for HDPE. Ethylene and butene-1 are blown into a tall vessel, together with the catalyst, which acts like a fluid. The polymers form quickly and drop out of the reactor bottom, where they can be easily cleaned up.

A liquid phase system also has been developed that uses the octene-1 comonomer.

Polypropylene

When polypropylene (PP) technology finally ripened in the late 1950s, the chemical industry was quick to harvest numerous applications. The primary attractions of this thermoplastic are the ease of molding or extruding it and its ability to hold color. Some familiar applications are automotive parts, luggage, pipe, bottles, fiber (particularly rope), housewares, and toys.

Some special problems arise in explaining polypropylene, and they breed a new vocabulary that is used throughout discussions on polymers more complicated than polyethylene. The problem lies in the extra group that propylene carries. Except for that methyl group ($-CH_3$), propylene would be ethylene. More of a problem, however, are the *allylic* hydrogens on the methyl group.* These hydrogens are reactive and easily displaced (see Fig. 19–5). That leads to branching and sometimes cross-linking, which can affect polymer properties. In fact, the difficulty

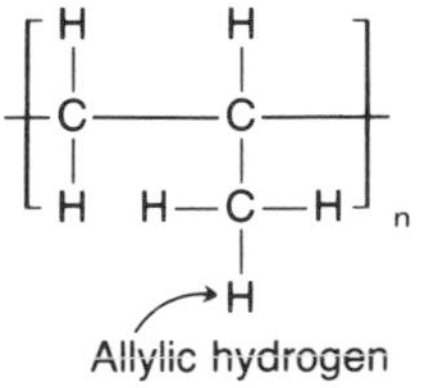

Allylic hydrogen

Fig. 19–5 Polypropylene

---

*Allylic comes from the Latin *allium,* meaning "garlic," C_3H_5 radical compounds are found in garlic and mustard.

of controlling branching and cross-linking slowed the commercial development of polypropylene until 1952. Then an Italian chemist, Giulio Natta, used Karl Ziegler's catalyst to produce a propylene polymer with some useful properties.

Understanding polypropylene chemistry requires knowing the critical difference between PP and the polyethylenes—the asymmetry of the molecular backbone. In polyethylene, every carbon looks like every other carbon in the chain. In polypropylene, the polymer linkage is between succeeding double-bonded carbons, like polyethylene. However, the methyl group survives as a branch on every second carbon in the backbone chain (see Fig. 19–6). Furthermore, branch orientation is crucial to the properties of the polymer.

There's a whole area of chemistry that deals with the spatial configurations of organic molecules called stereochemistry. In this field, you must have molecules with an asymmetrical carbon atom. That's one that has four dissimilar atoms or groups attached to it. Polypropylene has this condition on a repeating basis—the methyl groups are on every other backbone carbon. Such a polymer can be stereoregular or stereospecific.

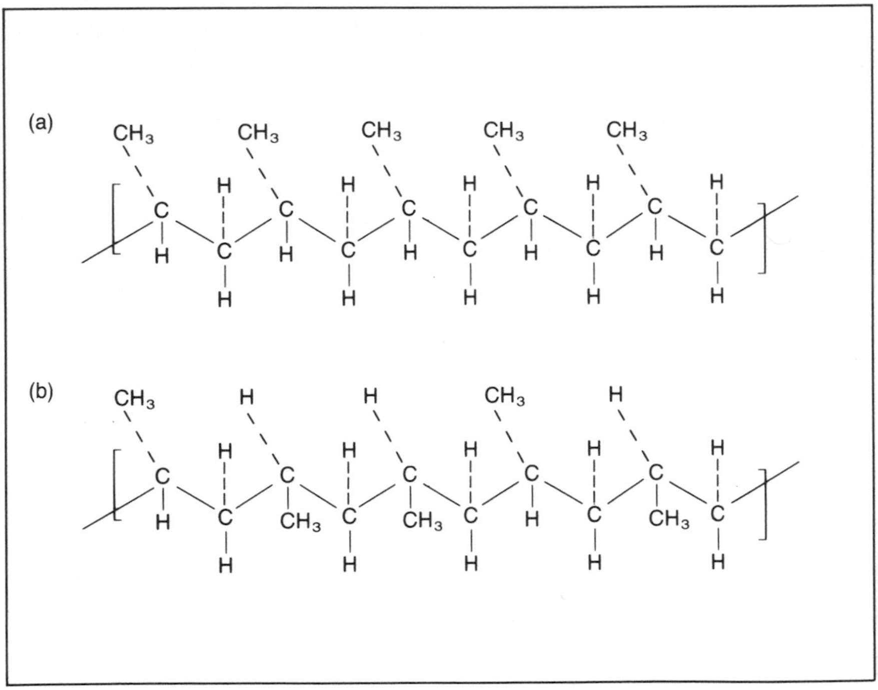

Fig. 19–6　(a) Isotatic polypropylene (methyl radicals in the same plane);
　　　　　(b) atactic polypropylene (methyl radicals randomly in and out)

In polypropylene, stereoregularity of the methyl group is important. It really makes a difference whether every one sticks out in the same direction (more accurately, in the same plane). There are three possibilities that have been identified in PP molecules:

Isotactic—all methyl groups are in the same plane;
Syndiotactic—methyl groups are alternately in the same plane;
Atactic—methyl groups are randomly in and out of the plane.

Isotactic polypropylene makes the best plastic. Atactic PP is soft, elastic, and rubbery, but it is not as good as either natural or synthetic rubber. The isotactic form has a high degree of crystallinity with the chains packed closer together, since the molecules are more regularly oriented. The greater crystallinity gives greater tensile strength, heat resistance, dimensional stability, and hardness and a high melting point.

So you can see why branching in the polymerization process can be a problem—the symmetry is affected. And you can get a hint why polypropylene was commercialized long after polyethylene. The chemistry and catalysis are much more demanding. That's why Giulio Natta won the Nobel Prize for his contribution.

Polypropylene Plant Description

The facilities for making polypropylene are similar to those for HDPE shown in Figs. 19–2 and 19–3. In fact, the same plants can be used. Like the stringent specifications for polyethylene plant feeds, propylene must be polymer grade (99.5% or higher). Water, oxygen, carbon monoxide, or carbon dioxide will poison the Ziegler-Natta catalyst.

The only difference from the HDPE and polypropylene processing steps is the addition of facilities to remove atactic and low molecular-weight PP from the isotactic, higher molecular-weight polymer. This is done by making a slurry of the polymer mixture in hot normal heptane, which will dissolve the low molecular-weight polymers and atactic PP. The isotactic does not dissolve and can be removed from the slurry by centrifuging.

Process improvements are still underway, mainly in the catalyst area. The objective is to increase the percentage of isotactic polypropylene, minimizing the atactic form. Although most operations are now producing 60–70% isotactic, some of the newer catalysts give almost 95%.

Polyvinyl Chloride

Raincoats, phonograph records, automobile upholstery, drainage pipe, roofing, and siding are all made from polyvinyl chloride (PVC). PVC is *the* consumer product plastic. Vinyl chloride polymers and copolymers are often referred to as vinyl resins. Originally they were based on acetylene. The switch to ethylene chemistry followed the development of the oxychlorination process for vinyl chloride described in chapter VIII. Today few acetylene-based VCM processing plants remain.

Polyvinyl chloride is the most important member of the vinyl resin family, which includes polyvinyl acetate (PVAC), polyvinyl alcohol (PVA), polyvinylidine chloride (PVdC) and polyvinyl acetal. Usually the term PVC includes polymers of VCM as well as copolymers that are mostly VCM.

Plasticizers

Polyvinyl chloride (the homopolymer) is rarely used alone. Usually additives and plasticizers are added, more so than for any of the other major thermoplastics. Plasticizers can change pure PVC from a tough, horny, rigid material to a soft, rubberlike form. Tricresyl phosphate (TCP, a gasoline additive in the 1960s) was the popular plasticizer, but dioctyl phthalate has since replaced it. Dioctyl phthalate is the primary end-use for 2-ethyl hexanol.

The plasticizer is usually added to the polymer during the compounding stage; that is, when it's being readied for molding, extruding, or rolling. The plasticizer is added in a hot mixer or roller operation. If PVC is to be plasticized, the polymerization steps can be controlled to produce a very porous polymer particle. Typically for a flexible PVC, 25–30% of the finished weight is plasticizer.

Adding a plasticizer such as dioctyl phthalate is called mechanical plasticization. Permanent or chemical plasticization can be done by copolymerizing VCM with monomers such as vinyl acetate, vinylidine chloride, methyl acrylate, or methacrylate. Comonomer levels vary from 5–40%. The purpose of the copolymers, of course, is to change certain properties such as softening point, thermal stability, flexibility, tensile strength, and solubility.

Another method to vary PVC properties is to add other polymers such as acrylonitrile-butadiene-styrene (ABS), styrene-acrylonitrile (SAN), methyl methacrylate (MMA), and nitrile rubber. The mechanical mixtures improve the processibility and the impact resistance of the rigid PVC products.

Manufacturing Polyvinyl Chloride

Like polypropylene, polyvinyl chloride has the problem of stereospecificity. The carbon atom to which the chloride atom is attached

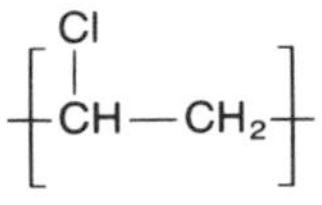

$$\left[\begin{matrix} & Cl \\ & | \\ -CH & -CH_2- \end{matrix}\right]$$

Fig. 19-7   Polyvinyl chloride

is asymmetric (Fig. 19-7). As a result, PVC molecules can be isotactic, syndiotactic, and atactic. Commercial PVC is only 5-10% crystalline. However, it is more dense (1.3-1.8 g/cc) than the polyolefins.

Vinyl chloride monomers can be polymerized by four processes: suspension, emulsion, bulk, and solution. Most PVC is made in the U.S. by the suspension method because the molecules are more suitable for molding, extruding, or calendering (that's calender, with an -*er*, which means "rolling into thin sheets").

Polyvinyl chloride in a latex form comes from the emulsion process, the second largest method. PVC latex can be used as is for coatings, or it can be processed for molding. In that case, it is spray dried, and the PVC particles are placed in a liquid plasticizer (called *plastisol*) or into a mixture of plasticizer and organic solvent (*organisol*). The PVC particles do not dissolve but remain dispersed until the mixture is heated. Then fusion occurs, yielding the final plastic object. This useful process forms special shapes by loading a mold with the plastisol or organisol and heating it.

PVC Plant Description

In the suspension polymerization process, the autoclave reactor is filled with water. Polyvinyl alcohol is the agent that helps stabilize the suspension. Lauroyl peroxide is the free radical catalyst that starts it all off. The reaction temperature is around 130°F, and the process takes 10-12 hours per batch with 95% conversion.

The reactors are typically 5,000-6,000 gallon, glass-lined, water-jacketed vessels (see Fig. 19-8). When the ingredients are loaded in vessels, steam is run through the jacket to raise the temperature to 120-150°F. After the reaction begins, cooling water replaces the steam in the jacket to take away the heat generated in the exothermic process. Meanwhile, the vessel contents are mixed vigorously to keep the monomer suspended in the water. The polyvinyl alcohol also helps out here. The polymer molecules must bump into each other to keep growing.

In order to get a porous PVC bead that will accept high levels of plasticizer, a sudden pressure release is sometimes used. During the 10-12 hours of cooking, the pressure increases as the temperature rises. Then the pressure declines slowly as polymerization approaches 100%.

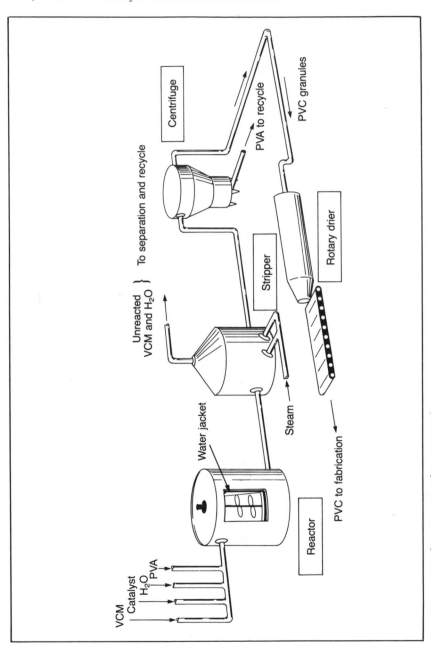

Fig. 19–8 PVC plant (suspension process)

If the pressure is suddenly released during the process, some VCM will vaporize. The granules of PVC that have formed begin to swell. Then as the vessel is "buttoned back up" and repressured, the VCM reliquefies, but the PVC remains swollen and porous.

When the reaction is completed, the suspension is transferred to a degassing tank, where steam strips out the unreacted vinyl chloride monomer. The PVC, now in a slurry with polyvinyl alcohol, is separated by centrifuging and is dried. The PVC powder or granules are then ready for additives and plasticizers, and can be fabricated into one of three mediums: calendered, extruded, or molded products.

Polystyrene

When you hear polystyrene (PS) you probably think of products made of PS foam: disposable coffee cups, packing materials, buoys, boat bumpers, and cheap ice chests. As a matter of fact, PS foam products are so important that there's a special section at the end of this chapter dealing with foam.

Foam accounts for less than half of the polystyrene output. The remaining products have properties very different from foam. Polystyrene is an excellent plastic for molded automobile and refrigerator parts. It accept color very well, so it is widely used in molding applications to simulate wood. Probably all the "wood" on your new console TV is made of polystyrene.

There's a lot of competition between PS and the other four thermoplastics, LDPE, HDPE, PP, and PVC. Polystyrene continues to lose market share, but it seems to have a permanent place in some applications, particularly molded foams (for carry-out food containers), some extrusions, and sheet and film applications.

Manufacturing Polystyrene

As with polypropylene and polyvinyl chloride, each repeating monomer unit in polystyrene has an asymmetric carbon atom (Fig. 19–9). The phenyl group (benzene ring) is attached to this carbon atom to make the polymer asymmetrical. The polymer can be iso-, syndio-, or atactic. Commercially produced PS is usually an atactic amorphous polymer (low crystallinity with good optics). The isotactic form can be

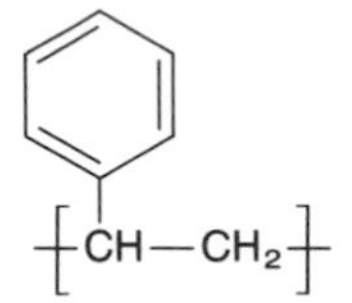

Fig. 19-9 Polystyrene

made using the Ziegler-type catalysts. However, there's no major improvement in the properties with these catalysts, so most processes produce the cheaper atactic.

All four polymerization processes can be used to make polystyrene. The reaction is an addition polymerization, using a free radical initiator (benzoyl peroxide or di-tertiary butyl peroxide). Usually the suspension or bulk process is used. The suspension process is identical to the PVC process shown in Fig. 19-8. As just one more mind expander, the bulk process should be discussed.

Polystyrene Plant Description

The bulk polymerization process needs monomers that can dissolve their own polymers. (There's no solvent or water in the reactor to keep the polymer floating around.) Polystyrene and some of its copolymers have this property, and so it's generally cheaper to use bulk polymerization.

The process begins in a *prepolymerizer*, which is a water-jacketed reactor with a mixer (Fig. 19-10). The styrene is partially polymerized by adding the peroxide initiator and heating to 240–250°F for about 4 hours. Approximately 30% of the styrene polymerizes, and the reactor contents become a syrupy goo. That's about as far as the prepolymer step can go—30% conversion—because the mixing and heat transfer get very inefficient as the goo thickens. Then the polymerization is difficult to control.

The goo is pumped to the top of a vertical, jacketed tower with internal temperature-regulating coils. The vessel is kept full of the styrene/PS mixture. A temperature gradient (change) of 280°F at the top to 400°F at the bottom also is maintained. The temperatures are controlled to prevent runaway reaction conditions and to permit 95% conversion of styrene to PS. As the polystyrene molecules grow, they sink to the bottom of the vessel and can be drawn off. The residence time in this vessel is 3–4 hours. The molten polystyrene is extruded to strands and is chopped into pellets before it is bagged.

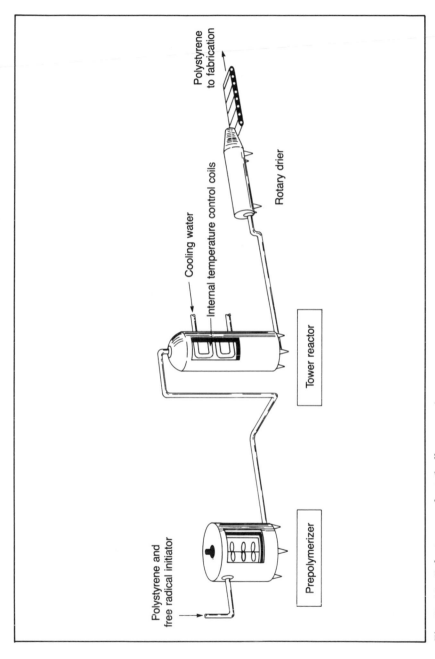

Fig. 19–10 Polystyrene plant (bulk process)

The most critical factor in this process is temperature control in the second reactor. The viscosity of the mixture from top to bottom changes with temperature and with PS concentration. If hot spots develop because of the exothermic reaction, the system can exceed optimum conditions. In that event, the batch must be quenched and, therefore, is ruined. Several process improvements (not shown in Fig. 19–10) include using agitators and small amounts of solvent.

Applications

About 25% of the polymerized styrene is in copolymer form. The largest volume copolymer is SBR (25% styrene and 75% butadiene), used to make tires, hoses, belts, and adhesives.

ABS (30% acrylonitrile, 20% butadiene, and 50% styrene) is a tough plastic with outstanding mechanical properties. ABS is one of the few plastics that combines both toughness and hardness. Applications include ballpoint pen shells, fishing boxes, extruded pipes, and space-vehicle mechanical parts. Approximately 20 pounds of ABS molded parts are used in one automobile.

SAN (70% styrene and 30% acrylonitrile) has better heat and chemical resistance and is stiffer than PS. However, the optical clarity is not as good. SAN is used in various houseware applications, particularly those items that contact food (chemical attack) and those that end up in a dishwasher (heat attack). Coffee pots and throw-away tableware are good examples.

Foams

Foamed polymers are low-density, cellular materials that contain bubbles of gas. They are made in various ways out of thermoplastics and thermosets. Their properties vary from rigid to flexible. The rigid foams are best known for their insulation properties (e.g., ice chests). The flexible foams are used extensively in cushioning (seats and mattresses).

The difference between rigid and flexible foams, from a simplified view, is the nature of the cells that make up the foam. Rigid foams are made of closed cells, and the gas they contain is sealed in. Flexible foams have open cells. When you compress a flexible foam, all of the air can be squeezed out. With a rigid foam the gas cannot escape, so nothing moves. Simple, isn't it?

The closed cells also give rigid foams excellent insulating properties. Gas is a notoriously poor heat conductor. That's why storm windows work. They have a dead air space in the middle that inhibits heat movement in or out. Rigid foams are just like storm windows. They trap dead air space. Flexible foams do not work as well because they let the air move around.

Foams are commercially produced several ways. Some polymerization processes produce their own foam. Polyurethanes, for example, are very exothermic. When they are formed in the presence of water, carbon dioxide is a byproduct. As the polymer forms, the CO_2 causes closed cell foam. As another example, a blowing agent can be injected into the molten polymer. The agent later decomposes into a gas such as nitrogen when the polymer is heated to melting. Epoxy resins also are expanded into foams in this manner.

A popular, related technique is to inject a volatile material into the polymer while it's still molten, causing it to foam immediately. For example, fluorocarbons are injected into polyurethane foam used for insulation. The trapped fluorocarbon is better than air as an insulator. Fluorocarbon or air also is used to expand PS foams. The gas is injected as the molten polymer is forced through a die. The foamed PS is then immediately injected into a mold to make items like egg cartons and trays for meat, produce, or fast-foods.

Expandable polystyrene beads are a material devised to accommodate the transportation drawbacks of foams. Foams take up a lot of room, but they are not heavy. Therefore, a truck or box car cannot be used very efficiently. Expandable PS beads can be readily turned into foam at the destination. The beads are impregnated with a volatile liquid like pentane as they are extruded, chopped, and cooled. Later on site, the beads are heated in small batches with steam. The vaporization temperature of the pentane is just below the melting point of the PS beads. As the beads soften, the pentane flashes and causes the polystyrene to foam. The polymer is then ready for molding. Coffee cups, ice chests, life preservers, buoys, and floats are often fabricated in this manner.

Most thermoplastics and thermosets can be foamed, many of them into either flexible or rigid form. The choice is controlled by the blowing agents, additives, surfactants, and mechanical handling processes. Some polymers can be expanded as much as 40 times their original density and still retain a substantial part of their strength. Most commercial foams are expanded to densities of 2–5 pounds per cubic foot. (Water density is 62 pounds.)

Plastic Properties

That worn-out joke applies: "The three most important things about plastics are properties, properties, and properties." It's impractical to cover all of the dimensions, but here are some of the most important.

Polyethylene has excellent electrical properties, good clarity, and good impact strength, and it is translucent in thick sections. It also has good chemical resistance and excellent processability.

Polypropylene is the lowest density plastic. It has fair-to-good impact strength and excellent colorability. It's translucent in thick sections, and it also has good chemical resistance. The properties can vary widely with different degrees of crystallinity. Polypropylene resists heat and has low water absorption. These characteristics make it a suitable material for many medical instruments that need sterilization by steaming.

Polyvinyl chloride has good electrical properties and is flame resistant with the proper plasticizers. It's even self-extinguishing. It has good impact strength and chemical resistance. Although it is produced in rigid form, it can be made flexible by adding plasticizers. It does require heat and light stabilizer additives.

Polystyrene is easily processible and has excellent color, transparency, and rigidity. it also has good dimensional stability, tensile strength, and electrical properties.

There, that's something nice to say about everything. But it still leaves you without a good grasp of which plastic to use for what application. As a last chance effort, you can look at Table 19–1. There's more information than you can possibly remember, but it's a good reference for future use.

TABLE 19–1
Thermoplastics—Uses and Properties

Thermoplastic	Typical use	Advantages	Limitations
LDPE	Packaging film, housewares, toys, wire insulation	Low cost, good optics, chemical resistance, flexibility, good impact strengh, dielectrics, and processibility	Low tensile strength, poor weathering, low stiffness, resists bonding and printing
HDPE	Blow-molded bottles, wire insulation, molded toys, pipe, packaging film, housewares	Same as LDPE but more rigid, tougher	Same as LDPE

TABLE 19–1 Continued
Thermoplastics—Uses and Properties

Thermoplastic	Typical Use	Advantages	Limitations
PP	Fibers, film, wire insulation, housewares, molded parts, medical wares	Lightweight, processibility, moderate tensile strength, rigid, tough, heat resistant, abrasion resistant, good dielectrics, low water absorption	Poor impact strength, embrittles below 0°F, poor weathering
PVC, flexible	Upholstery, molded products, wire coating, packaging film, grass bags	Excellent flexibility, flame resistant, chemical and abrasion resistant, self-extinguishing	Stiffens at low temperatures, hardness, embrittles with age (if plasticizers migrate)
PVC, rigid	Piping, siding, flooring, bottles	Tough; chemical, flame, and abrasion resistant; stiffness, self-extinguishing	Difficult to process, narrow temperature range
PS	Packaging, housewares, toys, appliance parts, foam, lighting fixtures	Clear, hard, easily molded, easily extruded, good tensile strength, stiffness, low cost, excellent electricals	Brittle, poor impact strength, poor solvent and chemical resistance, poor weathering, poor heat resistance, flammable

XX.

RESINS AND FIBERS

Tell her to make me a cambric shirt.

"Scarborough Fair" (1966)
Paul Simon and Art Garfunkel

After you've plowed through the thermoplastics, you only need to read about the resins and fibers to cover the rest of the applications for most petrochemicals. That's one reason for putting resins and fibers in one chapter. The other is that some polymers, like nylon, can be both a resin and a fiber. You just grow them a little differently.

The coverage in this chapter is compact—no detailed process descriptions or diagrams. Resins and fibers aren't really petrochemicals anyway. They're just a good climax to the petrochemical story.

Resins

Chapter XVIII didn't give you a very satisfying definition of resins. But it's useful here to talk about two classes of polymers called resins: thermosets and engineering thermoplastics.

Thermosets

You'll recall that thermosets are polymers that have lots of cross-linking. The molecules are three-dimensional, rather than two. More importantly once the cross-linking bonds are in place, the polymer becomes rigid and hard. Put another way, once the thermoset occurs, it is irreversibly set. That's the difference between thermosets and thermoplastics. The latter can be remolded, the former cannot. When you sweep up the scrap material around the molding/extruding machines that handle thermosets, you throw it away.

Phenolic Resins. The oldest condensation reaction on record is between phenol and formaldehyde to produce *phenolics*. The reaction was first documented by Baeyer in 1872. Thirty years later a technical application of this reaction was worked out by Dr. Leo Bakeland, when he showed that useful moldings can be made by carrying out the final stages of the reaction under pressure. As his reward, phenolic resins are still often called "Bakelite." They have become the workhorse of the plastics industry.

The chemistry of phenolic resins is complex. Even today it is not fully understood. The brief description that follows is definitely an over-simplification of the reactions involved.

To make phenolic resins, you must first make one of the two types of phenolic prepolymers: *resols* or *novalacs*. Resols are made using a base catalyst; novalacs, using an acid catalyst. Both are low molecular-weight, linear polymers that can be further reacted to give the cross-linked phenolic resin.

The reactions for resols are shown in abbreviated form in Fig. 20–1. Phenol and formaldehyde react in the presence of a base catalyst to form mono-, di-, and tri-substituted methylol phenols. These mole-cules further react at slightly higher temperatures to form the resol. Curing (cross-linking) the resol is done by adding heat. The links are primarily methylene, $—CH_2—$, and are three-dimensional.

The process for novalacs uses an acid catalyst and excess phenol. The reaction goes through the methylolphenol intermediate, too. How-ever, unlike the resols, novalacs do not have methylol groups (CH_2OH). Therefore they cannot cross-link with more heating. To get cross-linking, a little $(CH_2)_6N_4$ (hexamethylene tetramine or HMT) is added. When HMT is heated in the presence of trace amounts of water, some formal-dehyde and ammonia are generated, which provide the necessary con-ditions to add methylol groups. These molecules then cross-link with more heat to form the thermoset phenolic, just like the last step with the resols.

Phenolic resins are the cheapest of all molding materials, since they usually contain more than 50% filler—sawdust, glass fibers, oils, etc. Their main properties are heat resistance, excellent dielectrics, and ease of molding. However, they have poor impact resistance (they crack), and they don't hold most dyes very well, except black. Their use is thereby restricted—they're functional but not pretty. When the tele-phone companies started making phones in colors, they quit using phenolic resins and instead bought more expensive thermosets.

Other applications for phenolics are switchgears, handles, and ap-pliance parts, such as washing machine agitators (that's why they're usually black). Phenolics are widely used to bond plywood, particularly exterior and marine grades. Although urea-formaldehyde resins are cheaper for this purpose, they are not nearly as water resistant and have been limited to interior grades. Abrasive wheels and brake linings also are bonded with phenolic adhesives.

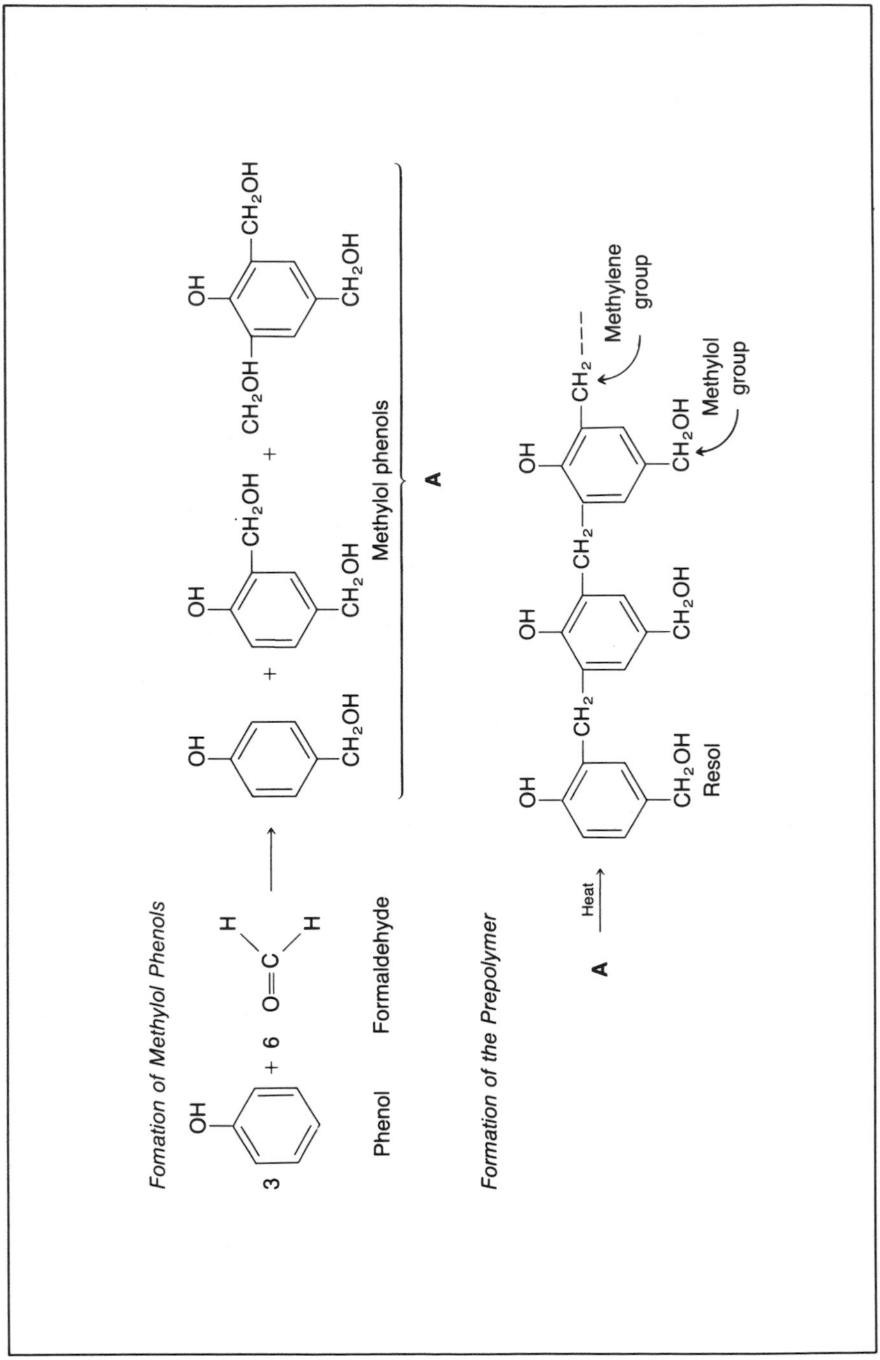

Fig. 20-1. Making phenolic resin via a resol

Fig. 20-1 continued

Resols and novalacs usually are used for molding in a powder form. After the prepolymer is formed (which is still a thermoplastic), the resin is solidified and is ground up. Then it is mixed with the filler, coloring, and in the case of the novalac, HMT. The final fabrication is done by pouring the powder into a mold and heating to the cross-linking temperature. Novalacs have better flow characteristics in the powder form and usually are preferred.

Resols usually are chosen for adhesives applications, especially plywood. They are used in the liquid prepolymer form. Plywood is formed by laying resol-soaked gauze in between the wood sheets and curing the gauze with pressure and temperature.

Epoxy Resins. The reaction of bisphenol A and epichlorohydrin gives a low molecular-weight linear polymer. This polymer further reacts with an amine curing agent, $-NH_2$, to give a general purpose thermoset. Now that's not as complicated as it sounds. First of all, you'd think the name epoxy is used because the epoxy ring, $-C\overset{\displaystyle O}{\overbrace{\qquad}}C-$, encountered in ethylene oxide is in the molecule. Well it is and it isn't. It is in the prepolymer at the ends of the polymer chain, which results from the bisphenol A/epichlorohydrin condensation reaction (Fig. 20–2). However, it is used up in the subsequent cross-linking step. This ring is the primary site where the polymer cross-links. A secondary site is the hydroxyl group that forms during the initial prepolymer condensation step.

Epoxy resins are a post-World War II development. Initially, they were used as surface coatings. They are extremely resistant to heat and corrosion, and they have excellent adhesion to metals. Unfortunately, they get chalky when exposed to too much sunlight. Therefore, their use is limited to primers and places protected from sunlight. Coatings are the largest volume use of epoxy resins, but the best known is the household adhesive that comes in two tubes. One contains the resin; the other has the amine catalyst and filler. Common amine curing agents are diethylene triamine (DETA) and triethylene tetramine (TETA). The package is usually labeled "epoxy glue."

Other major uses are commercial adhesives, laminates, and potting for electrical components. Epoxy resins are particularly suitable for potting (imbedding electrical components in a nonconductive thermoset) because they provide dimensional stability (no shrinking) as the thermoset cures.

Polyurethanes. The word urethane has the same root word as urine, because both are related to urea, NH_2CONH_2. All three chem-

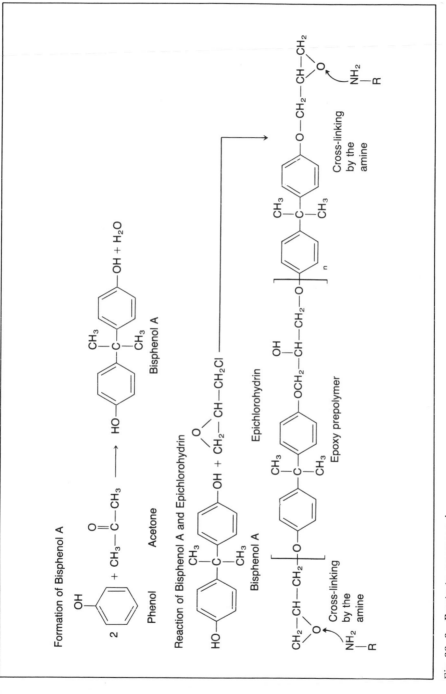

Fig. 20–2   Route to epoxy resin

icals have the characteristic —N—C— linkage. Urethanes result from
$$\begin{array}{cc} | & \| \\ H & O \end{array}$$
reactions between alcohols and isocyanates. So what's an isocyanate? That's a double-bonded compound of carbon, nitrogen, and oxygen, R—N=C=O. This compound is the product of an aliphatic or aromatic amine and phosgene. Remember the amine shown in Fig. 1–22? It's an R—NH$_2$. Phosgene is COCl$_2$, a product of the reaction of carbon monoxide and chlorine. Therefore, all of those elements go into making isocyanates.

Simple polyurethanes are made from *diisocyanates* and *diols*. The general form for a diisocyanate is O=C=N—**R**—N=C=O. The popular one, toluene diisocyanate (TDI), is shown in Fig. 20–3. Diols are molecules that have two hydroxyl group (alcohol signatures) attached, such as propylene glycol. Polymerizing TDI and propylene glycol gives a linear polymer because both molecules are bifunctional. That is, they each have only two sites where they can react, so they string out end to end.

If cross-linking is desired, as in a polyurethane foam, TDI can be

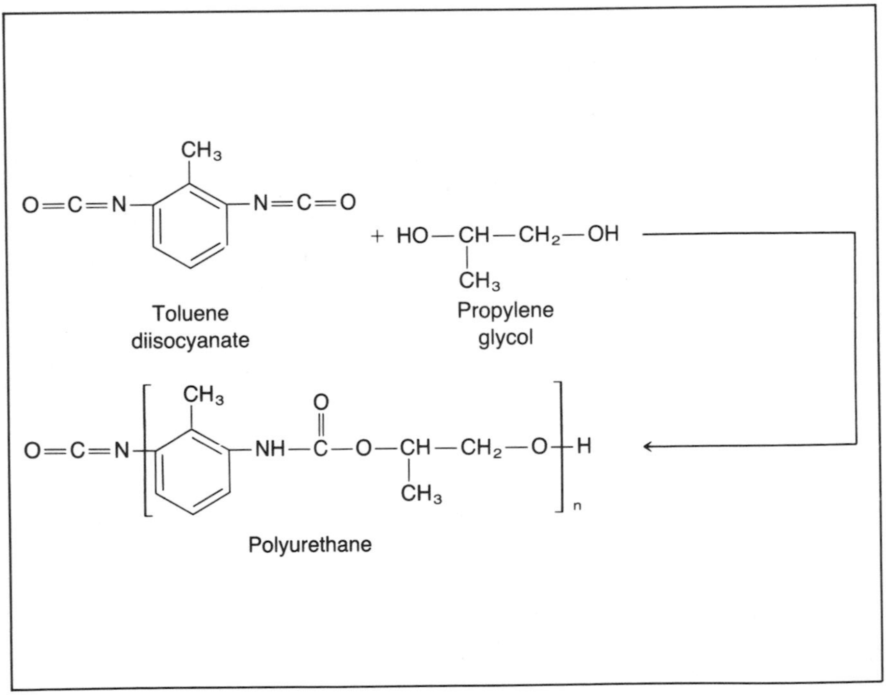

Fig. 20–3 Simple polyurethane process

reacted with a *polyol* or *polyolether*. These chemicals have multiple —OH groups that make them polyfunctional. Cross-linking can take place at one or more of the hydroxyl sites. Compounds other than TDI can be used to make polyurethanes. However, TDI accounts for about 60% of this production.

Polyurethanes are primarily found in rubbery foams, but they also make rigid foams and laminates. Two methods are used to develop these foams. One procedure mixes all of the ingredients, including a catalyst and a blowing agent additive, and pours them into a mold or void. The reaction kicks off immediately. Foaming is completed in a few minutes because the gases are liberated from the blowing agent. However, complete curing (cross-linking) takes several more hours.

The other foaming process is referred to as the *prepolymer* method. The monomers are reacted to form a low molecular-weight prepolymer. Later the prepolymer is mixed with small amounts of water and is heated. The water reacts with the free isocyanate groups to liberate carbon dioxide, which foams the polyurethane as cross-linking starts.

Polyurethane foams are lighter than foam rubber and have displaced many of its applications, such as in bedding, cushions, car seats, arm rests, and crash pads. Laminates are used widely in clothing as padding.

Rigid foams are excellent insulators, even better than polystyrene, and are used in refrigerators and refrigerated trucks and box cars. Polyurethane coating materials are popular additives to marine finishes and varnishes, particularly for gymnasium floors, bar tops, and other surfaces that take an abusive, abrasive beating.

Amino Resins. Urea, the first recorded, synthetically produced, organic compound, can be reacted with formaldehyde to form polymers called *urea-formaldehyde resins.* As with phenolic resins, their commercialization preceded an understanding of polymer chemistry. Melamine, the heterocyclic compound in Fig. 20–4, became an item of commerce in 1940 when a low-cost manufacturing route to resins was developed, based on urea. Melamine can be reacted with formaldehyde to form melamine-formaldehyde resins. The chemistry of both polymers, called amino resins, is similar to the phenolic resins.

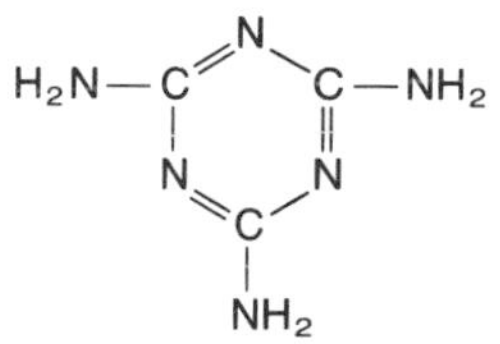

Fig. 20–4 Melamine, a heterocyclic

Articles made from these resins are water clear, hard, and strong, but they can crack. They have good electrical properties, and they have better colorability than phenolic resins. Amino resins are used as adhesives for plywood and particle board but only in interior grades. They have low weather resistance and deteriorate when exposed to sun, heat, cold, and moisture.

Extensive use of these resins is found in textile and paper treating and surface coatings. Many types of clothing also can be given a permanent press by melamine-formaldehyde treatment. Amino resins can be molded and are used for radio cabinets, buttons, and switch-plate covers. The melamine resins have a little better chemical and heat resistance and are used to make dishware and formica.

Engineering Resins

Polymers with special properties (such as high thermal stability, good chemical and weather resistance, transparency, self-lubrication, and good electrical properties) can be called engineering resins. Nylon and polycarbonates are two good examples.

Nylon. The name nylon covers a number of polymer compounds, all of which are based on the amide linkage $-\underset{\underset{O}{\|}}{C}-\underset{\underset{H}{|}}{N}-$ shown in Fig. 20–5. (Resins made from nylon actually account for much less volume than fibers from nylon.) The two most popular nylons, both in resins and fibers, are Nylon 6 and 66. These two account for about 80% of the production.

Nylon 6 is made by the addition polymerization of caprolactam. Caprolactam is a seven-sided heterocyclic that should be drawn as a septagon, just like cyclohexane and benzene are hexagons. However, few people can draw septagons well, so most sketch caprolactam as a rectangle, like the one shown in Fig. 20–6.

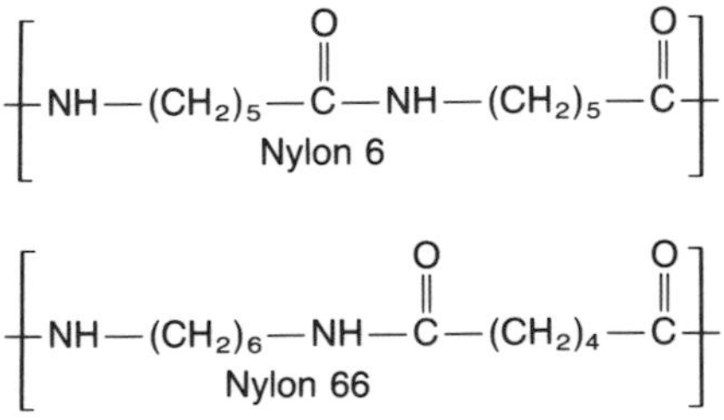

$$\left[-NH-(CH_2)_5-\overset{O}{\overset{\|}{C}}-NH-(CH_2)_5-\overset{O}{\overset{\|}{C}}-\right]$$

Nylon 6

$$\left[-NH-(CH_2)_6-NH-\overset{O}{\overset{\|}{C}}-(CH_2)_4-\overset{O}{\overset{\|}{C}}-\right]$$

Nylon 66

Fig. 20–5 Nylon

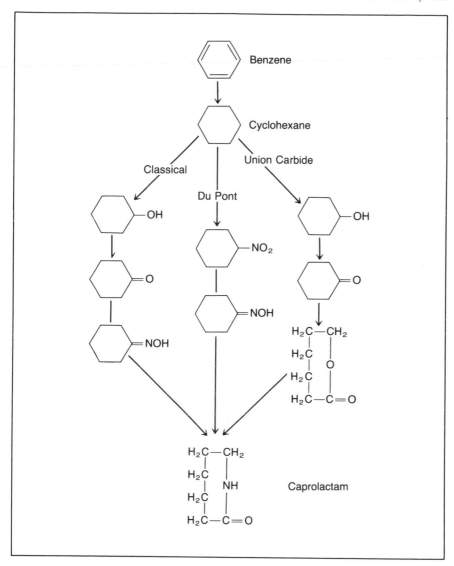

Fig. 20–6 Routes to caprolactam

There are several ways to make caprolactam. The most popular methods are shown in Fig. 20–6. They all begin with cyclohexane. (Or if you want to go way back, reread chapter IV where cyclohexane was made out of benzene.

Nylon 66 is made by the condensation polymerization of adipic acid and HMD (hexamethylenediamine). Adipic acid was covered in chapter XV, and HMD shouldn't be such a threatening word to you by now. The hex is six; the methylene is —CH_2—. The di is two, and amine is the signature group —NH_2 in Fig. 1–22. Together, these groups spell $H_2N(CH_2)_6NH_2$, which is HMD. The routes to HMD and adipic acid are shown in Fig. 20–7.

The 6 and 66 are part of an awkward numbering system used to indicate what a nylon was made from. A number up to 12 indicates the nylon was made from a single monomer with that number of carbons in it. A number over 12 signifies that two different monomers were used and the carbon counts in each. Check Fig. 20–5 to see how it works. Examples of other nylons are Nylon 11, 12, and 610. Originally all double numbered nylons had a comma in the name—Nylon 6,6. Through constant misuse, the comma has mostly been dropped.

Polycarbonates. The polycarbonates surfaced in the 1950s, so they are relatively new polymers. They are made in a condensation polymerization process. The reactants are either bisphenol A and phosgene or bisphenol A, phosgene, and phenol. Since bisphenol A is a derivative of phenol, the building block is the same in either case—phenol. The polycarbonate based on bisphenol A has the best balance of properties.

Polycarbonates differ mechanically from the epoxy resins they resemble chemically because polycarbonates are plastics. They can be

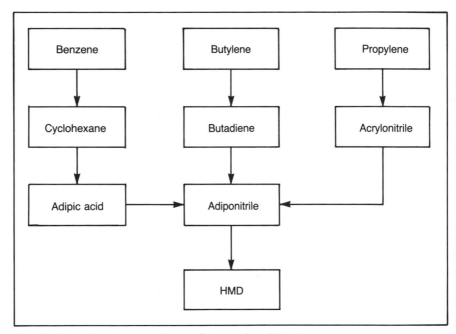

Fig. 20–7 Methods to produce adipic acid and HMD

molded or extruded from chips or crumbs. The products have high impact strength and can be used to make sturdy cast-films and transparent forms. Applications are photographic film; goggles; helmets; streetlight fixtures; tool, machine, and appliance casings; and electrical and electronic components.

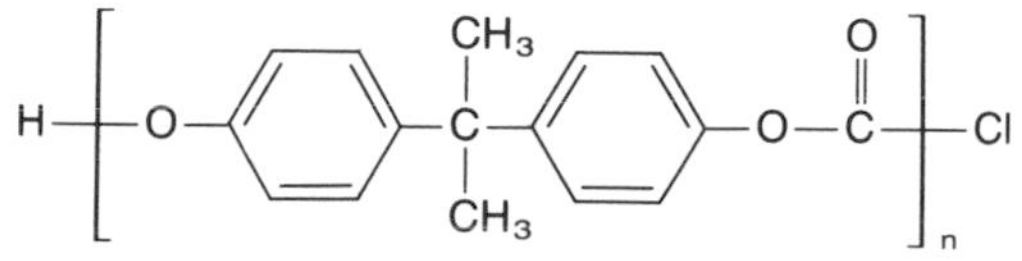

Fig. 20–8   Polycarbonate

Fibers

Natural fibers go back to prehistoric days. Probably one of the early applications was the conversion of a fiber (possibly wool or cellulose) into thread or rope strong enough to be used in a snare, net, or cage. Literature as far back as the 17th century notes that people attempted to make fibers out of something other than cotton, wool, or flax. The first manmade fiber, known as artificial silk, was made in the 19th century, when wood pulp was treated with nitric acid. The result was known chemically as cellulose nitrate and (eventually) commercially as Rayon. The commercial name referred to the sheen which "has the brilliance of the sun."

However, Rayon isn't a petrochemical derivative. Those products did not emerge until the technological breakthrough of W.H. Carothers of Du Pont in the 1930s. His classical research showed that chemicals of low molecular weight could be reacted to form polymers of high molecular weight. Some people consider Carothers the father of polymer chemistry. His original offspring was Nylon 66, the first commercial fiber to be made entirely from a synthetic polymer. (Some reports have PVC fiber beating Nylon 66 by two years, but PVC never really caught on as a fiber.)

The first Nylon 66 fiber production facility came on stream in 1939. The first nylon stockings were marketed in May 1940—just in time for G.I.'s to take them to Europe in World War II.

Du Pont continued their leadership role in synthetic fibers by commercializing acrylic fibers (Orlon) in 1950. They did a repeat performance in 1953 with polyester (Dacron). The big four fibers—Nylon 6, Nylon 66, acrylics, and polyester—now account for most of the synthetic production and about half of the fiber production of all kinds, including cotton, silk, and wool.

The Mechanics of Fibers

The chemistry of fibers is the same as that for resins. The important difference is the mechanics. For polymers to be suitable for fibers, you must be able to draw them into a fibrous form, normally by extrusion. Second, the size and shape of the molecules that make up the fiber must be correct. To have acceptable fiber properties, the molecules must be long, so they can be oriented to lie parallel to the axis of the fiber. Normally that's done (or enhanced) by drawing or stretching the fiber to several times its original length. The essential differences then, between resins and fibers, are the shape and the orientation of the molecules.

Spinning is the classic process of twisting a bundle of parallel short pieces of natural fiber into thread. (You can't get more classic than a spinning wheel.) It also is the modern process for extruding long lengths of manmade fibers. In the classic process, the short lengths are known as *staple* fibers and the resulting thread is a *spun yarn*. The long, extruded lengths are called *continuous filament* or just filament yarn.

The *denier* of a fiber or a yarn defines its linear density. This term describes the weight in grams of a 9,000 meter length at 70°F and 65% relative humidity. Denier generally denotes the size or diameter of the filament or yarn. Fibers usually range from 1–15 denier and yarns 15–1,650 denier. *Breaking tenacity* is a measure of the strength of a fiber or yarn and is measured in grams per denier. The stronger the fiber is, the higher the tenacity.

Nylon (Polyamide Fibers)

The chemical structure of the nylon fiber looks just like the nylon resin. The polymerization processes are the same; the numbering systems are the same; and the two most important nylon fibers are the same Nylon 6 and Nylon 66. The difference is the length of the molecule in comparison to the cross section. That's regulated by the polymerization process conditions.

After the polymer is formed, the filament is produced by extrusion. The nylon can be taken directly from the polymerization process while it is still molten. Alternatively, previously dried nylon chips can be heated and melted. The molten nylon is pressed through a *spinneret*, which is like a shower head with 10–100 holes. (A spinneret with a single hole produces a monofilament. Nylon monofilaments are used for fishing line and sheer hosiery.) The fine strands that form are immediately air-cooled, stretched, and wound on bobbins.

Then the fibers undergo *cold drawing*. Nylon 66 can be stretched 400–600% of its original length with very little effort. A very important process goes on during the stretch. The polymer molecules orient themselves to a controlled amount of crystallization, depending on the stretch conditions. This crystallization gives the fibers their special properties.

Finally the fibers are textured for specific applications. They can be twisted, coiled, or even randomly kinked, if they are to be used for carpet piling. More than half of the total carpet-fiber market is based on nylon staple and filaments. When nylon is to be blended with other fibers, the filament is cut into staple fiber—short pieces 3–15 centimeters in length—for subsequent spinning.

One of the largest uses of nylon fibers is tire cord. In apparel applications, which are another major area, permanent press can be achieved by heat treatment. This crease resistance lasts until abrasion, heat, or pressure wears down the molecule orientation. Since it is strong and lightweight, nylon also is used for rope, parachutes, and some undergarments.

Polyesters

The process for producing polyesters wasn't covered in the sections on thermoplastics or on resins. So read it carefully and it will make sense, despite the formidable nomenclature and diagrams. It's really not that hard.

Polyester is the Madison Avenue name for polyethylene terephthalate. The abbreviation is PET. Be careful, though, because PET is not a polyethylene-type chemical. It's a poly (ethylene terephthalate). If you remember that esters usually end in *ate*, then you can see that the "ester" in polyester is ethylene terephthalate.

Early production of PET involved the condensation polymerization of dimethyl terephthalate (DMT) and excess ethylene glycol (EG) in the presence of a catalyst. This process produced a low molecular-weight polyester prepolymer. The final step is to lengthen the prepolymer by heating it to about 525°F under a high vacuum, producing the final polyester. A simplified version of this reaction is shown in Fig. 20–9.

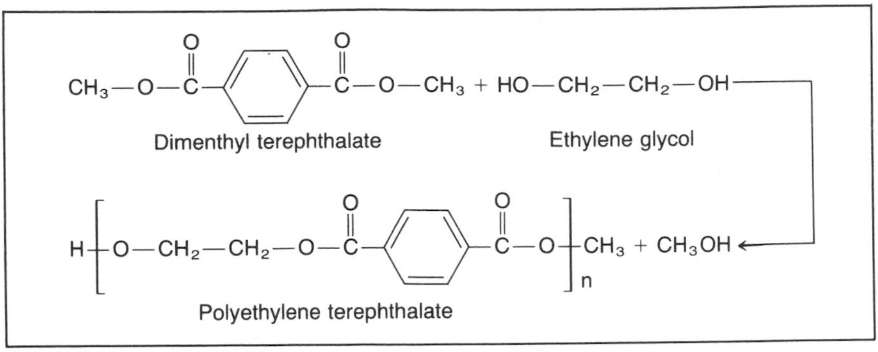

Fig. 20–9   Route to polyester

Like nylon, PET can be extruded, cooled, cut into chips, and stored for later *melt spinning*. It also can go directly to the spinneret. The downstream operations of the spinneret are much the same as with nylon.

A more direct route to polyester, using terephthalic acid (TPA) and bypassing DMT, was later developed in the 1960s. It wasn't until then that methods for purifying the TPA to fiber grade were developed and catalysts were found.

Unlike nylon which is highly crystalline, PET fibers are amorphous after spinning. They are like the molecules shown at the top of Fig. 18–9. In order to make a usable textile yarn or staple fiber out of PET, it must be drawn under conditions that result in orientation and crystallinity. This is accomplished by drawing at temperatures of about 175°F with stretch of 300–400%. As with nylon, the conditions of draw (especially the amount) determine the tensile strength and shrinkage properties. Industrial PET fibers, such as those suitable for tires, are more highly drawn.

Other important properties of PET fibers are heat setting (permanent press) and the ability to blend with cotton and wool. By appropriate texturing, various finishes are possible: fur-like deep pile for coats, jackets, bath mats, and soft toys and strong, coarse fiber for tire cords, V-belts, fire hoses, and carpeting. Polyesters are very resistant to degradation by sunlight. Dacron is still the most popular sail cloth because of its weight, its quick drying characteristic, its resistance to stretch and to mold, and its color fastness.

Acrylic Fibers.   Acrylic fibers are polymers of acrylonitrile and another chemical. When acrylonitrile is 85% or more of the polymer, the fiber is called acrylic. If there's more copolymer so the percentage of acrylonitrile decreases to 35–85%, the fiber is called *modacrylic.* Some of the popular monomers used as copolymers are methyl acrylate and methacrylate, acrylamide, vinyl acetate, vinylidene chloride, and vinyl chloride. Dynel is 40% acrylo and 60% vinyl chloride.

These polymers are generally made by the solution polymerization method. The cooking time in the reactor takes 30–60 minutes before the chains are long enough. The basic polyacrylonitrile chain is like the polypropylene chain—it grows along the ethylene backbone. The cyanide grouping is hung over the side.

Melt spinning is not used for polyacrylics because they are sensitive to high temperatures. They actually begin to decompose before they reach a melting temperature. Solution spinning is used instead. The dried polymer is dissolved in a polar solvent like acetone or dimethyl-formamide (DMF). The spinning mechanics are otherwise the same, except the solvent is recovered as it vaporizes, immediately after the extrusion through the spinneret. Most acrylics are sold and used in the form of staple fiber.

Acrylic fibers generally have good "hand," as it's called in the business (they're soft). They resist creasing, and they are quick to dry. Acrylics have replaced wool in many applications, such as blankets and sweaters. Because of their unique *bulking* characteristics, they take on the appearance of wool yarn.

Polypropylene Fibers. A small part of the total fibers market (and therefore of this section on fibers) is fiber grade polypropylene. The chemistry for polypropylene fibers is the same as for thermoplastics. The spinning mechanics are the same as that for nylon. Polypropylene fibers are particularly resistant to abrasion and chemicals, and they are lightweight. However, they don't take colors very well, and the materials have low softening points and low resilience (they wrinkle). The major applications for polypropylene fibers are carpet-face fiber and backing (because it's tough) and rope (because it is strong and floats in water).

I hate quotations. Tell me what you know.

Journal, Ralph Waldo Emerson
(1803–1882)

How do you summarize a summary of a whole body of chemistry like the polymers? You don't. You only pick out some tidbits.

The differences in the properties of polymers goes back to the chemical configurations. In simple terms, thermoplastics can be molded because they are long chain molecules that slip if pushed or pulled, especially at higher temperatures. Thermosets are cross-linked, so the long chains stay put under stress, strain, or heat. The fibers get their flexibility and strength when the polymer molecules align during filament formation.

Tables 20–1, 20–2, and 20–3 summarize the ingredients and the chemical structures of some of the more important thermoplastics, thermosets, resins, and fibers.

If you've gotten this far (even if you've skipped a couple of pages or chapters), you need to be reminded of the good news. There's an excellent index coming up right after the appendix. You can use it to refresh your memory on those one or two points that you might forget sometime in the future.

TABLE 20–1
Thermoplastics

Abbreviation	Name	Polymer form
LDPE	Low-density polyethylene	$-CH_2-$
HDPE	High-density polyethylene	$-CH_2-$
LLDPE	Linear low-density polyethylene	$-CH_2-$
PP	Polypropylene	$-CH_2-\overset{\displaystyle CH_3}{CH}-$
PVC	Polyvinyl chloride	$-CH_2-\overset{\displaystyle Cl}{CH}-$
PS	Polystyrene	$-CH_2-\overset{\displaystyle C_6H_5}{CH}-$

TABLE 20–2
Thermosets

Name	Ingredients	Polymer form before cross-linking
Phenolics	Phenol and formaldehyde	
Polyurethane	Diisocyanate and a polyglycol	$-O-R-O-C(=O)-NH-R'-NH-C(=O)-$
Epoxy resin	Bisphenol A and epichlorohydrin	

Urea-formaldehyde

Formaldehyde
and urea

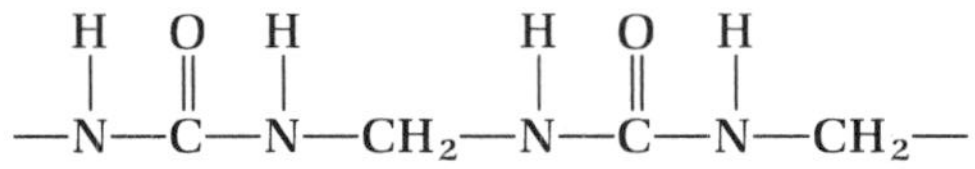

Melamine-formaldehyde

Formaldehyde
and melamine

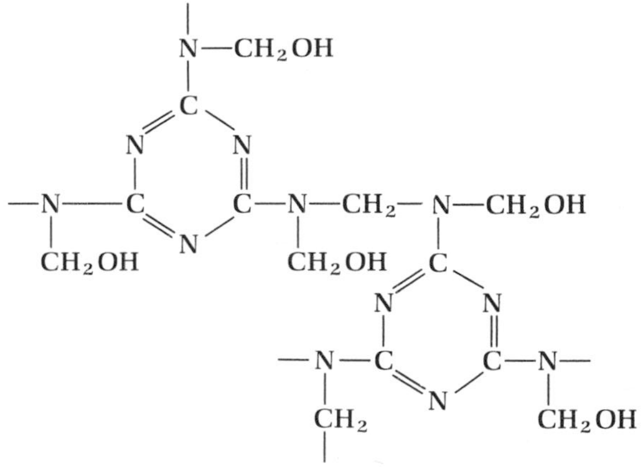

TABLE 20–3
Resins and Fibers

Name	Ingredients	Polymer form
Nylon 6	Caprolactam (from cyclohexane)	$-NH-(CH_2)_5-\overset{\displaystyle O}{\overset{\displaystyle \|}{C}}-$
Nylon 66	Adipic acid and hexamethylenediamine	$-NH-(CH_2)_6-NH-\overset{\displaystyle O}{\overset{\displaystyle \|}{C}}-(CH_2)_4-\overset{\displaystyle O}{\overset{\displaystyle \|}{C}}-$
Polycarbonate	Bisphenol A and Phosgene	
Polyacrylics	Acrylonitrile	$-\underset{\underset{\displaystyle N}{\overset{\displaystyle \|\|\|}{C}}}{CH}-CH_2-$
PET	Polyester	

APPENDIX
CONVERSION AND YIELD

Conversion and yield are often used to describe the efficiency of a plant or a process. Sometimes the semantics get mixed up and one is substituted erroneously for the other.

Conversion is defined as the percent of the feed that disappears in a chemical reaction. For example, take the pounds of benzene coming out of the ethylbenzene reactor and divide by the pounds of benzene going into the reactor. Subtract that number from 1.0 and multiply by 100. The answer is the percentage of benzene conversion. Conversion is usually measured around a reactor, rather than around the whole plant. For example in an ethylbenzene plant, virtually no benzene leaves the site because it is recycled. Therefore, *plant* conversion is often not a very helpful concept, but the conversion across the reactor is.

Yield is a more difficult concept because it's used in several ways and because one of the uses requires some basic chemistry. In the simplest case, yield refers to the pounds of product leaving a reactor divided by the pounds of feed. For example in an olefins plant, the ethylene yield is equal to the pounds of ethylene divided by the pounds of, say, gas oil feed. (This definition of yield is more commonly used in refining than in petrochemicals; however, ethylene plants are usually in refineries.)

The more technical petrochemical definition of yield refers to the amount of *converted* feed that actually ends up as product. (This is not an easy subject for the "nontechnical" person, but here goes.)

Definitions

Atomic weight. The relative weight of an atom. Hydrogen is the lightest and is defined to be 1. Carbon is 12 and oxygen is 16, meaning they are 12 and 16 times as heavy as hydrogen.

Molecular weight. The sum of the (relative) weights that make up a molecule. The molecular weight of water, H_2O, is $2 + 16 = 18$; for methane, CH_4, it is $12 + 4 = 16$.

Mole. The molecular weight of a compound expressed in grams is the *gram molecular weight,* or *mole.* If the weight is expressed in kilograms or pounds, it's called kilogram moles or pound moles. A mole of water is 18 grams of water; a pound mole of methane is 16 pounds of methane.

The theoretical yield of a reaction can be calculated based on molecular weights. For example, the reaction of hydrogen with carbon dioxide to produce methanol is as follows:

$$3H_2 \quad + \quad CO_2 \quad \longrightarrow$$
$$3(1 + 1) + (12 + 16 + 16)$$

Molecular weights: 6   +   44

$$CH_3OH \quad\quad + \quad H_2O$$
$$(12 + 1 + 1 + 1 + 16 + 1) + (1 + 1 + 16)$$

Molecular weights: 32   +   18

Theoretically, three moles of hydrogen (6 grams) react with one mole of carbon dioxide (44 grams) to give one mole of methanol (32 grams) and one mole of water (18 grams). In actual practice a byproduct, dimethyl ether, also forms:

$$2CH_3OH \quad \longrightarrow \quad CH_3OCH_3 + H_2O$$

In an example experiment, 22 grams of carbon dioxide are reacted with excess hydrogen to give 12 grams of methanol and some dimethyl ether. To determine the *actual yield* of methanol based on the carbon dioxide reacted, follow this procedure:

Step 1. Determine the moles of CO_2 reacted and the moles of CH_3OH produced.
22 grams of CO_2 is half of a mole of CO_2 (22/44)
12 grams of CH_3OH is 0.375 mole (12/32)

Step 2. Theoretically, one mole of CO_2 gives one mole of CH_3OH; therefore, half of a mole of CO_2 gives half of a mole of CH_3OH.

Step 3. Divide the actual moles of product produced by the theoretical moles of product produced:
$$\frac{0.375}{0.500} = 0.75$$
The actual yield is 75%.

Note that the yield is not the weight of the product (12 grams) divided by the weight of the feed (22 grams). Using molecular weights as an adjustment helps track what happens chemically to the modecules, so that the chemist or chemical engineer can determine whether to work on improving the process.

ABOUT THE AUTHORS

Donald L. Burdick is a graduate of Creighton University and Kansas University. He received a doctorate in organic chemistry from Kansas in 1957. After a short period in the academic world, Burdick pursued careers in industrial research and later in marketing. Most of his experience is in the field of petrochemicals, and he is presently associated with a major oil company. Burdick has co-authored several scientific papers and has twelve patents credited to his name.

William L. Leffler received a bachelor of science degree in industrial management from M.I.T. and masters and doctorate degrees in business administration from the Business School of New York University. He has written numerous papers for management and technical journals, primarily in the petroleum industry. A major oil company employee, Leffler also is the author of *The Economics and Technology of the U.S. Propane Industry* (Pennwell, 1973) and *Petroleum Refining for the Nontechnical Person* (Pennwell, 1979).

INDEX